HERE AND THERE:
A FIRE SURVEY

To the Last Smoke

SERIES BY STEPHEN J. PYNE

STEPHEN J. PYNE

HERE AND THERE

A Fire Survey

THE UNIVERSITY OF
ARIZONA PRESS

TUCSON

The University of Arizona Press
www.uapress.arizona.edu

© 2018 by The Arizona Board of Regents
All rights reserved. Published 2018

ISBN-13: 978-0-8165-3853-9 (paper)

Cover design by Leigh McDonald
Cover art: *Lesnoi pozhar* by A. K. Denisov-Uralsky [detail]

Library of Congress Cataloging-in-Publication Data are available at the Library of Congress.

Printed in the United States of America
♾ This paper meets the requirements of ANSI/NISO Z39.48-1992 (Permanence of Paper).

To Sonja
old flame, eternal flame

৽

CONTENTS

SERIES PREFACE
To the Last Smoke

WHEN I DETERMINED to write the fire history of America in recent times, I conceived the project in two voices. One was the narrative voice of a play-by-play announcer. *Between Two Fires: A Fire History of Contemporary America* would relate what happened, when, where, and to and by whom. Because of its scope it pivoted around ideas and institutions, and its major characters were fires or fire seasons. It viewed the American fire scene from the perspective of a surveillance satellite.

The other voice was that of a color commentator. I called it *To the Last Smoke*, and it would poke around in the pixels and polygons of particular practices, places, and persons. My original belief was that it would assume the form of an anthology of essays and would match the narrative play-by-play in bulk. But that didn't happen. Instead the essays proliferated and began to self-organize by regions.

I began with the major hearths of American fire, where a fire culture gave a distinctive hue to fire practices. That pointed to Florida, California, and the Northern Rockies, and to that oft-overlooked hearth around the Flint Hills of the Great Plains. I added the Southwest because that was the region I knew best. The interior West beckoned because I thought I knew its central theme and wanted to learn more about its margins. Alaska boasted its own regional subculture. Then there were stray essays that needed to be corralled into a volume, and there were all those relevant regions that needed at least token treatment. Some, like the Lake States and Northeast, no longer commanded the national scene as they once had, but their stories were interesting and needed recording, or, like the Pacific Northwest or central oak woodlands, spoke to the evolution of fire's American century in a new way. I would include as many as possible into a grand suite of short books.

My original title now referred to that suite, not to a single volume, but I kept it because it seemed appropriate and because it resonated with my own relationship to fire. I began my career as a smokechaser on the North Rim of Grand Canyon in 1967. That was the last year the National Park Service hewed to the 10 a.m. policy and we rookies were enjoined to stay with every fire until "the last smoke" was out. By the time the series appears, 50 years will have passed since that inaugural summer. I no longer fight fire; I long ago traded in my pulaski for a pencil. But I have continued to engage it with mind and heart, and this unique survey of regional pyrogeography is my way of staying with it to the end.

Funding for the project came from the U.S. Forest Service, Department of the Interior, and Joint Fire Science Program. I'm grateful to them all for their support. The University of Arizona Press deserves praise as well as thanks for seeing the resulting texts into print. And I'd like to acknowledge Kerry Smith for his meticulous copyediting, which has once again saved me from my worst grammatical self.

PREFACE TO VOLUME 9

M OST OF THESE ESSAYS grew out of travels while I researched
what became *Between Two Fires* and the other volumes of *To
the Last Smoke*. Not everything I found and learned fit into a
regional framework because some topics were grounded in ideas or disciplines, not places. A few essays resulted from conferences or workshops,
or spun off from field trips or stray thoughts. A handful found publication
in specialized journals or as op-ed pieces not likely to be read by the fire
community. A few were written before this specific project and are dated
in their newsiness, but I intended the whole series to serve as a cross sec-
tion of the American fire scene as viewed by a consistent voice and vision,
and felt they contributed. The group under "Fire Far" describes scenes
outside the United States. They are gathered here because I find that we
know things by context, by contrast and comparison, and because, in my
mind, these pieces are the final surge of a wave train that began nearly
a decade ago and went as far inland as my energies and talents allowed.

ACKNOWLEDGMENTS

M OST OF THESE ESSAYS in this book, though not all, were written while I was researching a fire history of the United States with funding from the U.S. Forest Service, Department of the Interior, and Joint Fire Science Program. I'm grateful for their support. The following essays were first published in some form as indicated:

"The Cash-Value of Fire History: An Apologia," in *Proceedings, 3rd Fire Behavior and Fuels Conference*, International Association of Wildland Fire (October 2010), 1–4.

"Untamed Art," *Forest History Today* (Fall 2008), 48–57.

"Coming to a Forest Near You," as "Box and Burn: The New Reality of Western Fire," *Slate* (July 4, 2014), http://www.slate.com/articles /technology/future_tense/2014/07/box_and_burn_the_future_of_u _s_wildfire_policy.html.

"East of the Wind and West of the Rain," *ISLE: Interdisciplinary Studies in Literature and the Environment* (Winter 2015).

"The Wrath of Kuhn" as "Problems, Paradoxes, Paradigms: Triangulating Fire Research," *International Journal of Wildland Fire* 16 (2007), 271–76.

"Slouching Toward Gatlinburg," as "What the East Can Learn from the West in Firefighting," *History News Network* (December 7, 2016), http://historynewsnetwork.org/article/164582.

"Portal to the Pyrocene," as "Welcome to the Pyrocene," *Slate* (May 16, 2016), http://www.slate.com/articles/technology/future_tensc/2016/05/the_fort_mcmurray_fire_climate_change_and_our_fossil_fuel_powered_society.html.

"Smokey's Cubs," *History News Network* (September 13, 2015), http://historynewsnetwork.org/article/160530.

"In the Line of Fire: An Overview of Wildfire in Korea," *Forest History Today* (Spring/Fall 2015), 50–53.

"Burning Banff," *ISLE* 11.1 (Summer 2004), 221–47.

HERE AND THERE:
A FIRE SURVEY

FIRE NEAR

A CACHE OF ESSAYS that speak to fire in the American scene, though not grounded in any of the regions that *To the Last Smoke* has surveyed. Some are themes, some are topics, a few are categorical orphans. But the American fire scene has ever resisted simple pigeonholing.

OUR COMING FIRE AGE

A Prolegomenon

In 2015 I gave a TED talk on "Fire, a Biography." I distilled the whole history of fire into 14 minutes. Here, I allow myself some more words and space for a useful digression or two. What follows is a thumbnail history of fire on Earth.

AT NIGHT, VIEWED FROM SPACE, the cluster of lights looks like a supernova erupting in North Dakota. The lights are as distinctive a feature of night-time North America as the glaring swathe of the northeast megalopolis. Less dense than those of Chicago, as expansive as those of Greater Atlanta, more coherent than the scattershot of illuminations that characterizes the Midwest and the South, the exploding array of lights define both a geographic patch and a distinctive era of Earth's history.

Nearly all the evening lights across the United States are electrical. But the constellation above North Dakota is made up of gas flares. Viewed up close, they resemble monstrous Bunsen burners, combusting excess natural gas released from fracking the Bakken shale. In 2014 the flares burned nearly a third of the fracked gas. They constitute one of the most distinctive features of the U.S. nightscape. We might call them the constellation Bakken.

While the flares rise upward, the fire front is actually burning downward into the outgassing drill holes as surely as a candle flame burns down its tallow stalk. The flames are descending as rapidly as their fuels are rising. They are burning through deep time, combusting lithic landscapes from the geologic past and releasing their effluent into a geologic future. Eerily, the Bakken shale dates to the Devonian, the era that records the first fossil charcoal, our first geologic record of burned material. Its gases will linger through the Anthropocene.

In 1860 the English scientist Michael Faraday published a series of public lectures in which he used a candle to illustrate the principles of natural philosophy. Fire was an apt exemplar because it integrates its surroundings, and it was apt, too, because in Faraday's world, fire was ever present. Every nook and cranny of the human world flickered with flames for lighting, heating, cooking, working, and even entertaining. But that was starting to change. By then, Britain had 10,000 miles of railways and the U.S. had 29,000. Those locomotives demanded more fuels than the living landscape could supply. Engineers turned to ancient landscapes—to fossil biomass, notably coal—and they simplified fire into combustion.

Today, a modern Faraday would not use a candle—probably couldn't because the lecture hall would be outfitted with smoke detectors and automated sprinklers, and his audience wouldn't relate to what they saw because they no longer have the lore of daily burning around them. For a contemporary equivalent he might well turn to a fracking flare, and to illustrate the principles behind Earthly dynamics he might track those flames as they burn down through the deep past of fire and humanity.

Earth, water, air—all are substances. Fire is a reaction. It synthesizes its surroundings, takes its character from its context. It burns one way in peat, another in tallgrass prairie, and yet another through lodgepole pine; it behaves differently in mountains than on plains; it burns hot and fast when the air is dry and breezy, and it might not burn at all in fog. It's a shapeshifter.

The intellectual idea of fire is a shapeshifter, too. In ancient times, fire had standing with the other elements as a foundational axiom of nature. In 1720, the Dutch botanist Herman Boerhaave could still declare that "if you make a mistake in your exposition of the Nature of Fire, your error will spread to all the branches of physics, and this is because, in all natural productions, Fire . . . is always the chief agent."

By the end of the 18th century, fire tumbled from its pedestal to begin a declining career as a subset of chemistry and thermodynamics, and a concern only of applied fields such as forestry. Fire no longer had intellectual integrity: it was considered a derivation from other, more fundamental principles. Just at the time open fire began retiring from quotidian life, so it began a long recession from the life of the mind.

Fire's fundamentals reside in the living world. Life created the oxygen fire needs; life created the fuels. The chemistry of fire is a biochemistry: fire takes apart what photosynthesis puts together. When it happens in cells, we call it respiration. When it occurs in the wide world, we call it fire.

As soon as plants colonized land in the Silurian period about 440 million years ago, they burned. They have burned ever since. Fires are older than pines, prairies, and insects. But nature's fires are patchy in space and time. Some places, some eras, burn routinely; others, episodically; and a few, only rarely. The basic rhythm is one of wetting and drying. A landscape has to be wet enough to grow combustibles, and dry enough, at least occasionally, to allow them to burn. Sand deserts don't burn because nothing grows; rainforests don't burn unless a dry spell leaches away moisture. Biomes rich in fine particles such as ferns, shrubs, and conifer needles can burn easily and briskly. Landscapes laden with peat or encumbered with large trunks burn poorly, and only when leveraged with drought.

As life and the atmosphere evolved, so did fire. When oxygen thickened in the Carboniferous period around 300 million years ago, dragonflies grew as large as seagulls, and fires swelled in like proportion such that 2 to 13 percent of the era's abundant coal beds consist of fossil charcoal. When grasses emerged in the Miocene, they lavished kindling that quickened fire's spread. When animals evolved to feast on those grasses, fire and herbivores had to compete because that same biomass was fodder for each: it could go down into gullets or up in flames but not both.

Today, ecologists refer to landscape fire as a disturbance akin to hurricanes or ice storms. It makes more sense to imagine fire as an ecological catalyst. Floods and windstorms can flourish without a particle of life present: fire cannot; it literally feeds off hydrocarbons. So as atmosphere and biosphere have changed, as oxygen has ebbed and flowed, as flora and fauna have sculpted biomass into new forms, so fire has evolved, morphing into the pyric equivalent of new species.

Yet there was one requirement for fire that escaped life's grasp, the spark of ignition that connected flame with fuel. Ignition relied on lightning, and lightning's lottery had its own logic. Then a creature emerged to rig the odds in favor of fire. Just when hominins acquired the capacity to manipulate fire is unknown. But we know that *Homo erectus* could tend fires and, by the advent of *Homo sapiens*, hominins could make fire at will.

A revolutionary phase change all around. Until that Promethean moment, fire history had remained a subset of natural history, particularly of climate history. Now, notch by notch, fire gradually ratcheted into a new era in which natural history, including climate, would become subsets of fire history. In a sense, the rhythms of anthropogenic fire began to replace the Milankovitch climate cycles, which had governed the coming and going of ice ages. A fire age was in the making.

Earth had a new source of ecological energy. Places that were prone to burn but had lacked regular ignition (think Mediterranean biomes) now got it, and places that burned more or less routinely had their fire rhythms tweaked to suit their human fire tenders. Species and biomes began a vast reshuffling that defined new winners and losers.

The species that won biggest was ourselves. Fire changed us, even to our genome. We got small guts and big heads because we could cook food. We went to the top of the food chain because we could cook landscapes. And we have become a geologic force because our fire technology has so evolved that we have begun to cook the planet. Our pact with fire made us what we are.

We hold fire as a species monopoly. We will not share it willingly with any other species. Other creatures knock over trees, dig holes in the ground, hunt—we do fire. It's our ecological signature. Our capture of fire is our first experiment with domestication, and it might may well be our first Faustian bargain.

Still, ignition came with limits. Not every spark will spread; not every fire will behave as we wish. We could repurpose fire to our own ends, but we could not conjure fire where nature would not allow it. Our firepower was limited by the receptivity of the land, an appreciation lodged in many fire-origin myths in which fire, once liberated, escapes into plants and stones and has to be coaxed out with effort.

Those limits began to fall away as people reworked the land to alter its combustibility. We could slash woods, drain peat, loose livestock—in a score of ways we could reconfigure the existing biota to increase its flammability. For fire history this is the essential meaning of agriculture, most of which, outside of floodplains, depends on the biotic jolt of burning to fumigate and fertilize. For a brief spell, the old vegetation is driven off, and a site is lush with ashy nutrients, and—temporarily—imported cultivars can flourish.

In 1954, the American anthropologist and nature writer Loren Eiseley likened humanity itself to a flame—spreading widely and transmuting whatever we touch. This process began with hunting and foraging practices, but sharpened with agriculture. Most of our domesticated crops and our domesticated livestock originate in fire-prone habitats, places prone to wet-dry cycles and so easily manipulated by fire-wielding humans. The way to colonize new lands was to burn them so that, for a while, they resembled the cultivars' landscapes of origin.

Yet again, there were limits. There was only so much that could be coaxed or coerced out of a place before it would degrade, and there were only so many new worlds to discover and colonize. If people wanted more firepower—and it seems that most of us always do—we would have to find another source of fuel. We found it by reaching into the deep past and exhuming lithic landscapes, the fossil fallow of an industrial society.

Instead of redirecting or expanding fire, the conversion to industrial burning removed open flame, simplified it into chemical combustion, and stuffed it into special chambers. Instead of being constrained by the abundance of fuels, anthropogenic fire was constrained by sinks, the capacity of land, air, and ocean to absorb its byproducts. The new combustion was no longer subject to the old ecological checks and balances. It could burn day and night, winter and summer, through drought and deluge. Its guiding rhythms were no longer wind, sun, and the seasons of growth and dormancy, but the cycles of human economies.

The transformation—call it the pyric transition—was as disruptive as the coming of aboriginal firestick and fire-catalyzed farming, but it was more massive, much faster, and far more damaging. Some landscapes burned to their roots. Seasonally, skies were smoke palls. Frontier settlements might vanish in flames. The pyric transition runs through fire history and Earth's pyrogeography like a terminator.

Eventually, as the new order prevailed, as it wiped flame away by technological substitution and outright suppression, the population of fires plummeted, leading to ecological fire famines. The transformation might have left Earth with too much generic combustion, and too much of its effluent lodged in the atmosphere, but the industrialized world also left too little of the right kind of fire where it's needed.

Promoting the steam engine developed with his business partner, James Watt, in the late 1770s, Matthew Boulton boasted to the biographer James Boswell that they sold what all the world wanted—power. In 1820, a year after Watt died, Percy Shelley published *Prometheus Unbound*, in which he celebrated the unshackling of the unrepentant Titan who had brought fire to humanity. By then, the use of coal, and later oil, was liberating a generation of New Prometheans.

This newly bestowed firepower came without traditional bounds. For a million years the problem before hominins had been to find more stuff to burn and to keep the flames bright. Now the problem became what to do with all the effluent of that burning and how to put flame back where it had been unwisely taken away.

The new energy revolution leveraged every activity, like fire itself creating the conditions for its spread, each reinforcing the other. But the collateral damage in the form of wrecked landscapes could not be ignored. Engineers sought to keep fire within the machines, not loosed on the countryside. Countries, particularly those with extensive frontiers, public lands, or colonial holdings, sought to shield their national estate from fire. They set lands aside to shelter them from promiscuous and abusive burning and sought to control fires when they occurred.

State-sponsored conservation had considerable currency among progressive thinkers. When Rudyard Kipling wrote "In the Rukh" (1893), a story that explained what became of *The Jungle Book*'s Mowgli after he grew up, he had him join the Indian Forest Department and fight against poaching and "jungle fires." Only later would the paradoxes become palpable. Only later would overseers realize how hard they would have to struggle to reinstate fire for its ecological benefits.

But the flames were only the visible edge of a planetary phase change. The slopover that followed once Earth's keystone species for fire changed its combustion habits is best known for destabilizing climate. But humanity's new firepower has a greater reach, and the knock-on effects are rippling through the planet's biosphere independently of global warming. The new energy is rewiring the ecological circuitry of the Earth. It has scrambled ecosystems and is replacing biodiversity with a mechanical pyrodiversity—a bestiary of machines run directly or indirectly from industrial combustion. The velocity and volume of change is so great that observers have begun to speak of a new geologic epoch, a successor to the

Pleistocene, that they call the Anthropocene. It might equally be called the Pyrocene. The Earth is shedding its cycle of ice ages for a fire age.

The traditional view of North Dakota, as of the Great Plains generally, divides it into humid east and arid west with the border between them running roughly along the 100th meridian. It's a division by water but it works for fire as well. It also marks a potential boundary between Pleistocene and Anthropocene.

For Pleistocene Dakota, look east to the prairie pothole region. It's a vestigial landscape of the ice sheets. The retreating ice left a surface dappled and rumpled with kettles, drumlins, eskers, potholes, kames, and ridges that slowly smoothed into a terrain of swales and uplands. The swales filled with water. The uplands sprouted tallgrass prairie. Those ponds make the region a vital flyway for North American waterfowl. But keeping the wetlands wet is only half the management issue. The birds nest and feed in the uplands and, being clothed in tallgrass prairie, the uplands flourish best when routinely burned. Few of these fires start from lightning; the only viable source is people, who followed the retreating ice and set fire to the grasses. Those fires are themselves relics of a bygone epoch. They annually renew the living landscape that succeeded the dead ice.

For Anthropocene Dakota, look west to the Bakken constellation. Not only is it a symbol of industrial combustion, but a major source of greenhouse gases and a catalyst for land use change and all the rest of the upsets and unhingings and scramblings that add up to make the Anthropocene. The flares speak to the extravagance of industrial fire—burning just to burn in order to get more stuff to burn. It's both a positive feedback and an eerily closed loop that accelerates the process and worsens its consequences. Instead of seasonal waterfowl, vehicles powered by internal combustion engines traverse the landscape ceaselessly.

East and west represent two kinds of fires and two kinds of future for humanity as keeper of the planetary flame. One is a Promethean narrative that speaks of fire as technological power, as something abstracted from its setting, perhaps by violence, certainly as something held in defiance of an existing order. The other is a more primeval narrative in which fire is a companion on our journey and part of a shared stewardship of the living world.

Sometime over the past century, we crossed the 100th meridian of Earth history and shed an ice age for a fire age. Landscape flames are yielding to combustion in chambers, and controlled burns, to feral fires. The more we burn, the more the Earth evolves to accept still further burning. It's a geologic inflection as powerful as the alignment of mountains, seas, and planetary wobbles that tilted the Pliocene into the cycle of ice ages that defines the Pleistocene.

The era of the ice is also our era. We are creatures of the Pleistocene as fully as mastodons and polar bears. Early hominins suffered extinctions along with so many other creatures as the tidal ice rose and fell. But humans found in the firestick an Archimedean fulcrum by which to leverage their will. For tens of millennia we used it within the framework bequeathed by the retreating ice, and for more than a century we have been told that we thrived only in a halcyon age, an interglacial, before the ice must inevitably return.

Gradually, however, that lever lengthened until, with industrial fire, we could unhinge even the climate and replace ice (with which we can do little) with fire (with which we can seemingly do everything). We can melt ice sheets. We can define geologic eras. It seems Eiseley was right. We are a flame.

COMING TO A
FOREST NEAR YOU

Looking back on the fire revolution, and ahead to the era that seems poised to replace it. For the West, box and burn may not be the best option around but is the only one that can serve as a foundational strategy.

WHEN CALIFORNIA LAST had a drought this severe, the United States was on the cusp of a reformation in how it managed fire on public wildlands. Fifteen years earlier an insurrection had boiled over into a full-fledged revolution that moved the federal agencies from a policy committed to fire's control to one dedicated to fire's management. They would restore a natural process to natural sites. They would reclaim humanity's oldest technology to its rightful place in working landscapes. Hot fire would reemerge as the newest of cool tools.

But the summer of 1977 was not quite the time nor California the right place. The U.S. Forest Service was still months away from officially publishing its reformed policy, and the National Park Service, building on a decade of experimentation, continued to fuss over a manual that would guide planning across its diverse holdings. Besides, the West was in drought; California's was brutal; even the two Sierra parks that had pioneered the national campaign to reinstate fire, Sequoia-Kings Canyon and Yosemite, were battening down for an anticipated siege. Now was not the time to play with fire.

The fires, when they came, were fought with every engine, hotshot crew, and airtanker the country could muster. In August I joined a

National Park Service crew drafted from the western parks to fight the Marble-Cone fire in the Ventana Wilderness outside Big Sur. Wilderness be damned, bulldozers scraped lines across ridgetops in a vain attempt to halt a mammoth blaze that threatened the watershed of Carmel. That fire continued, eventually becoming the second largest on record for California. Other fires followed. They didn't cease until the Honda fire overran Vandenberg Air Force Base during the winter solstice.

The season was a classic callout: the firefight on a subcontinental scale. It confirmed what the 1970 blowout had demonstrated, that no single agency could by itself cope with the big burns, that the future of fire protection depended on interagency sharing, or what was known as total mobility. It also pointed to what officials hoped would be the future of wildland fire, that fewer fires would need to be fought on this scale because we could let remote ones burn themselves out (and celebrate the biological work they did) while substituting our own quasi-tamed fires for wild ones. Before the next fire season arrived, the policies and institutional reforms to make that happen were in place. America's great cultural revolution on fire had arrived. Prescribed fire would broker between let-burning and suppression only.

Then the revolution stumbled.

In retrospect it's easy to see how America's western firescapes evolved into an alloy of the big, the costly, and the feral. How prescribed fire became too complex, too expensive, and too laden with liabilities. How concerns with firefighter safety pushed fire agencies to pull back under extreme conditions and hostile settings. How legacy fuels powered more savage outbreaks. How climate change removed the temporal buffers and exurban sprawl the spatial ones that had granted previous generations room for maneuver. How political gridlock over public lands paralyzed agencies like the Forest Service. How costs argued to go big with escape fires and multiple ignitions. How the only fires and smokes that were allowed were, paradoxically, wildfires for which there was no culpable agent. How, in sum, fire agencies have (if silently) ceded an illusion of control, have surrendered beliefs that they can get ahead of the problem, and have sought, with big-box burnouts and point protection, to turn necessity into opportunity.

Widen the historical aperture, and it is possible to see the present scene as a logical, though not inevitable, span in a narrative arc that began a century ago. For 50 years following the Big Blowup of 1910 the country, under the aegis of the U.S. Forest Service, had sought to remove fire. For the past 50 years, under a consortium of institutions, it has sought to restore fire as fully as possible, for the most part under a doctrine of fire by prescription. The first strategy only destabilized biota after biota. The second has failed to reinstate fire regimes on anything like the scale required. Both ambitions, however, have shared the belief voiced by William Greeley in 1911 that fire control was as surely a matter for scientific management as is silviculture.

Suppression had never been—could never be—as successful as advocates wished. Neither did restoration fail as utterly as critics might claim. In the Southeast prescribed fire has become a foundational practice, in principle on par with suppression. While no one, not even Florida, burns as much as they want or should, the practice is so accepted there that controversies turn on what season or mix of burns should be applied. In the West prescribed fire has succeeded in select grasslands and in prairies overrun with shrubs, but elsewhere it exists as a boutique practice or as pile burning. The belief so prominent during the early years of the 1960s fire revolution that tame fire might replace or complement natural fire has faltered. We aren't going to restore landscapes to a prior golden age or fashion new ones to our liking for a golden age to come. We can't classify fires by some rational schema that nature will recognize. We can't separate the various practices of fire management into bureaucratically correct categories. Science can't solve what are, fundamentally, political issues.

It's also easy to see where current trends are heading and what we might do to improve the outcomes. The emerging issue is how to cope with whatever wildland fires appear on the land. Their source of ignition doesn't matter—fire is fire, as the mantra goes. Nor does it matter much where they originate. Unless they threaten high-value assets like exurbs or sequoia groves, it makes little sense to dictate responses based on legal locations. There will never be enough money, enough political capital, or enough space for maneuvering to get ahead of the threats or to subject wildland fire to the categories of managerial theories. We will have to deal with what is coming at us.

―――――――

Instead, fire management has become a mash-up. A given fire might have parts fought, parts boosted by burnouts, parts loose-herded, parts monitored, parts burned one day and extinguished the next. The trend is clear. Suppression will concentrate on the point protection of critical assets whether the enclaves be ecological or exurban. Wildland fires will be confined within big boxes of landscape whose perimeters are then burned out. Area burned will rise. Costs per acre will decline. The West's wildlands will continue to cascade from one record year to another with very mixed outcomes. Some fraction of the burns will be more severe than we will be comfortable with. Some fraction within fire perimeters will remain unburned. The rest will exhibit burning more or less within what we might consider the slippery range of historic variability.

Such burns, and the burnouts that accompany them, will constitute the bulk of prescribed fire in the West, provided one ignores the inconvenient fact that they are not really prescribed. They just happen. They are a fusion of wild and set flames. They are not directed toward the prescriptions of a golden past or those of an imagined golden future. They offer resilience, not restoration. They implicitly acknowledge that humans, while fire's keystone species, are not in control. We can, within limits, start, stop, and shape fire and we can certainly disrupt fire regimes, but we will only truly be in charge within built landscapes; the more meticulously those places are shaped, the greater our grip. We have pretty rigorous control over what happens within a diesel piston. We have very little over free-burning flame in wildlands where our reach exceeds our grasp.

―――――――

Such developments do not mean we can do nothing or that research is irrelevant. But what might we do? And what topics merit systematic investigation?

Begin with burning out. On too many fires, the most severely damaged sites are those from backfires. We can do better. Particularly when firing occurs a ridge away from the main burn or from secure lines under less than emergency conditions, there are good reasons to consider these counterfires as a variant of prescribed burning. Surely our fire behavior knowledge is sufficient to prevent ruinous aftershocks. We could spotburn and chevron-burn rather than strip-fire. We could adjust burning according to what would be de facto prescriptions. The know-how and technology exist.

Those defensible perimeters can look a lot like the preattack zones of an earlier age. We could construct them in advance so that they meet not only considerations of fire behavior and firefighter safety but make sense in terms of ecological benefits: we could design them to be burned with consequences beyond their simple value as counterfires. We are already doing some of this with strategic treatments for fuels. Rationalize the process a bit further, and you have a matrix of burning blocks. When the fire revolution began, those blocks would have been the basis for prescribed fire. Now they must accept mixes of fires, a fusion of wild flames and burnouts. That mash-up is the future of fire management in the West. It's a hybrid fire—half prescribed, half suppressed.

Among large-perimeter fires, the interior will present a dappled texture of sites burned to their roots, of sites unburned, and of sites variously touched by flame. This is an ideal matrix for postburn or recovery firing. Some of those new burns and reburns will happen immediately; some, over a sequence of years. It sounds counterintuitive to argue for more fire as soon as the ash cools, but that postburn firescape is often a perfect setting in which to prescribe burn because it offers safe perimeters. You can burn against the black. You can burn in different seasons. You can burn, in patches, over large areas. Call the immediate process salvage burning. Call what occurs later prescribed reburns or recovery burns.

All this, too, is a suitable topic for research. Too often the failure to follow up on a wildfire has meant, particularly in montane forests, that one or two or more cycles may pass before fire returns in any form. Instead, we are presented with circumstances that could accept fire without the usual protocols of checklists and the gambler's-ruin logic of set piece burns. We need a large cache of burning options to match the various possibilities of the postburn landscape. Here science could advise, and then measure the outcomes.

Such a strategy is not restoration as that enterprise has traditionally been understood. It does not consciously seek to return the scene to a prior condition, nor does it shape it to a predetermined future state. It does not conform to that idealized project in which science determines and management applies. It makes do with what is at hand. It's a philosophy of resilience, not restoration. It's pragmatism with a small *p*.

From its origins federal fire programs have channeled their research energies into science and policy. But besides science we need poetry, and in addition to policy we need politics. The poetry is what gives meaning to the enterprise. The politics is what gives it legitimacy and institutional leverage. These topics are amenable to the scholarship known to history, literature, art, and philosophy. Just as we use science to stiffen practices in the field, so we can look to other disciplines to strengthen emerging notions of what is desirable and how to get there. Adaptive management, for example, is an avatar of Pragmatism, America's contribution to the philosophical tradition of Western civilization. That small *p* pragmatism we are witnessing in the field might well be improved with some upgraded logic from big *P* Pragmatism. Why not?

―――――――――――

As the folk saying goes, we can't choose what happens to us, but we can choose our response, and in the case of fire management, we have various practical and intellectual tools to give heft and rigor to the choices we might make. We can control very little of what is driving fire in our western wildlands. In reality, the whole preoccupation with drivers is misguided, because the better analogy to wildfire is a driverless car that barrels down the road integrating everything around it. At times one factor or another may dominate, but they are all present. Their very abundance means there are many points of intervention possible.

There is an adage about writing that reads, "Follow your heart but use your head." Logic and craft are what transform experience and intuition into art. A similar epigram will likely apply as we confront the fires to come. We will have to follow the flames—we'll have no choice—but we will need to use our head. We'll need everything we have if we wish to turn the chaotic mash-ups of fire's presence into an artful management of our public wildlands.

This time I won't be patrolling Chews Ridge, pulaski in hand, looking for places where sparks and errant flames have breached the line. I'll be looking for signs that something has crossed the fuelbreaks of history. We had 50 years of unremitting effort to resist all fires. We have then undergone 50 years trying to restore good fires. We now seem to be entering a more modest era of resilience. How we handle those spot fires and slopovers will tell whether the future will mean more force and counterforce, or whether we are finally learning to live with our best friend and worst enemy.

SCIENCE SUPPLIES
THE SOLUTION

Thoughts provoked by the National Academy workshop on fire research, and what science can and can't do in managing landscape fire.

THE ENTRY WAS AUSPICIOUS; the setting, august; and the times, desperate. The National Academy of Sciences had been established in 1863, amid the American Civil War, to bring scientific advice to the federal government. Its imposing building sits on Constitution Avenue, facing the Mall. Its logo features a torch. Its Great Hall spills into the auditorium beneath a mural that depicts Prometheus carrying his flame. For fire folk tasked with discussing the future of wildland fire science at a time when fires were hollowing out the U.S. Forest Service, it couldn't get more distinguished.

The Workshop on a Century of Wildland Fire Research, hosted by the academy on August 27, 2017, gathered many of the best minds in the field. The session took place against a political order in which a new president famous for not reading, who seems to govern by TV and tweets, whose spokespersons preferred "alternative facts" to vetted empiricism, whose temperament seemed more likely to wield a flamethrower than a hose. The country was not openly in arms with itself but was more implacably divided into warring tribes than at any time since the Civil War. Even the National Cohesive Strategy (NCS) segregated the country into three realms, which, revealingly, was also a map of Civil War America.

Chief Forester Tom Tidwell announced to the assemblage both its charge and his challenge. Wildland fire was becoming an existential crisis for the U.S. Forest Service—burning through budgets, disrupting landscapes, shredding plans, mocking the premise behind the agency's century-old charter to apply rational, science-based solutions to the national forests. These were dire times. The 2016 fires had absorbed over 50 percent of the Forest Service's entire budget, and there was no fiscal suture in sight. Deputy Chief Carlos Rodriguez-Franco noted that a science of wildland fire had begun, largely under the Forest Service, a century ago. Now was the time to inventory and data-mine that past. Now was the desperate moment to project the knowledge gained and find an answer that would quell the flames and stanch the hemorrhaging of monies that was bleeding the agency white.

Yet like the torch on the academy's logo, the founding faith of the agency still burned bright. The chief declared, as an axiom, that "science supplies the solution." What science are we missing? he asked. What gap in research, if filled, will allow us to complete our understanding? What elusive gadget of modern technology, found or devised, will permit us to translate that revealed knowledge into practice? What was the missing link of scientific expertise that would ignite the National Cohesive Strategy into a rational program to address the nation's escalating need to halt bad fire, promote good fire, and end the fiscal immolation of its agencies?

Implicit in the chief's plea was the recognition that the wildland-urban interface (WUI) had unbalanced the national fire scene. Its metastasizing spread over rural America had pulled the federal agencies away from land management and was hammering them into a national fire service that would, by endlessly suppressing fire, ultimately only worsen the scene. My mind drifted. The WUI was a true problem but a tired trope.

My thoughts finally alighted on what might well be regarded as the interface's founding fire—call it WUI One. I recalled images of burned houses and columns of fire engines. On the screen of my memory I watched snatches of *Design for Disaster*, the documentary produced by the Los Angeles Fire Department, and replayed its sonorous narration by William Conrad. Through the smoke of memory I saw two canonical images that 56 years later still spoke truth to the power of fire.

═══════════

One image was specific to California—call it California Iconic. The other was a testimony in twisted irony that could have occurred anywhere but happened to fall out this time in California. Call it California Fallout.

Iconic California is the template image, endlessly recycled by photo-journalists, of some doofus on a wood-shingle roof with a garden hose. Except that this time the doofus was Richard Nixon in white shirt and tie, vacantly staring up and away from the background flames while a piddling stream of water collected at his feet.

The usual interpretation is that the photo shows the indifference to hazard by Southern Californians, who then expect firefighting to plug the gap, or that politicians are ever eager for staged photo ops. But what I read in this version is that politics is not an afterthought when it comes to fire management. Wildland fire, particularly the WUI, is a matter of public safety and public assets that properly belongs in political discourse. It has its distortions and comic perversions—Nixon with a garden hose is a great illustration—but politics is a fundamental reality of the scene. The belief that fire can be managed by disinterested agents who apply the tested conclusions of science is an illusion. Politics has been there, right from WUI One.

California Fallout features another member of the great-and-famous. Willard Libby had, the year previously, been awarded a Nobel Prize in physics for his 1949 work on ^{14}carbon dating. In the intervening years he had joined the Atomic Energy Commission and become a public shill for fallout shelters. Two weeks before the Bel Air-Brentwood fire, he was photographed, in formal wear, complete with bow tie, in what he called a "poor-man's fallout shelter," which he had built in his backyard out of sand bags and railroad ties (along with a little plastic sheeting) for $30.

The fire incinerated his shelter along with his home. His wife fled the flames with her mink coat and his Nobel medallion. A repeat photo of the shelter taken after the burn, sans Libby, failed to garner the same interest. Learning of the episode, Leo Szilard declared it proved two things. One, God exists. And two, He has a sense of humor. It would also prove that fire was less predictable than the other effects of nukes.

Or more broadly, it hinted at the limitations of science to resolve the problems that would plague the WUI. Unlike blast and much of fallout, fire was a biological construct and a creature of context. It propagated. Its spread depended on wind, fuel moisture, the arrangement of combustibles, most of which grew according to ecological imperatives or, where built by people, by cultural considerations, all of which would likely be shattered by a bomb's blast. So, too, fire's character as friend or foe, as a transient summer job or a chronic bureaucratic crisis, depended on its social and political setting. In March 2017 that context seemed desperate.

The assembled scientists did their best. They talked, they traded thoughts, they proposed paths forward. But in the end, they testified, if by their deeper silence, to the limitations of science. There was no shame in that: it was the way of the world. The fire crisis was really a crisis in the American experiment itself. The wildfires burning through agencies were burning through the widening cracks in the country's culture and politics. What the Forest Service needed was not a scientific breakthrough or a problem-disrupting technology but a clarity of mission, a legitimacy before the many American publics and their representatives, an intellectual civility and a cultural humility. It didn't need to upgrade its science-informed policy. It needed a poet.

The chief was proposing a world that would not have Nixon on the roof or Libby in his shelter. That was not a dishonorable ambition, just a utopian one. Until then fire management officers on the ground would glance at the science as codified in decision support systems and play with new gadgets, but they would act as conditions permitted. They would try to respond in a way that was more than the sum of the parts that sent fire through the land. They couldn't hope to master fire. They could hope to listen better to what the fire was telling them, and if they were lucky as well as good, they could hope to stay ahead of the flames.

THE RED PRESCRIBED
LONGLEAF COCKADED WIRE
GRASS BURNING REFUGE

I was supposed to be writing Awful Splendour, *a fire history of Canada, which was what I said I would do when I was accepted for an academic year at the National Humanities Center. But Sandhills National Wildlife Refuge wasn't far away, and I knew it had a successful burning program, and I was ready to leave my monastic cell at the center for some time in the woods.*

WHEN THE FLAMES HIT the wire grass, the sun was slatting through early morning shadows, and the January air still held a nip of evening wetness, and the dribbled fire gulped and nestled amid the stalks, and then pushed out with a crackle. To both sides other flames formed a thickening line, thin-streaming out of a driptorch, three parts diesel to one part gasoline, a liquid fire that drizzled over grassy clumps, rust-colored pine needles, a splash of oak leaves, a patch of love grass, windfallen twigs and branches, a longleaf sapling like a green fountain spraying needles. The flames flashed steadily as they consumed the petro mix, then probed to the bio-fueled flanks, flaring in a patch of resinous grass, shrinking to gnaw a slab of bark, slowing through a thin-layered cast of needles, leaping from tussock to tussock, probing and testing and adjusting flame to fuel, yet overall becoming thicker, and taller, and moving with the light winds, steadying into a line of fire, advancing like a shallow broken surf over the tumbled land and around the big trunks, lofting a veil of smoke into the canopy. Twenty feet back, another line, staggered to this one, fought to reach its stride.

The time has come to fire off burn unit 15/02, one of more than a hundred such blocks in the Carolina Sandhills National Wildlife Refuge (NWR), and the first of the season. The burn will continue as long as combustibles remain and the many prescriptions governing how much smoke might billow forth will allow. It will continue until flame has passed over the entire block and left it a-smolder. Another block will follow, and another, as weather and oversight agencies permit, a jagged, gerrymandered mosaic of sites sprouting longleaf woods, oak scrublands, slash pine plantations, pocosins and switch cane, old fields and fresh-drilled winter wheatgrass, and the assorted biotic flotsam left from centuries of a hard-worked land. The fires will burn across a history as fractured as the landscape. They will persist for months, as opportunities allow. They will burn until roughly a third of the refuge's 45,000 acres have passed through the flames.

These fires are not optional. The longleaf pine, ideally, forms a savanna, the trees tall, the vista open, the rolling surfaces fluffed with wire grass, love grass, or other tussocky graminoids, chocked with forbs, a biota as species rich as any in North America. It's easy to burn such places: flames crackle briskly through the grasses, open to the wind, blackening the thick trunks, perhaps scorching the crown with killing heat. The trees, unless dead and dried, don't burn. The ease of burning reflects, in fact, a long history of burning, a chronicle measured not merely in fuel accumulation or in centuries, or even in millennia, but in evolutionary time.

The longleaf pine is one of the most fire-accommodating species in America. While still a seedling, it passes through a grassy stage in which it resembles the tufted clumps that surround it, and it burns just as they do. Out of this almost unavoidable immolation, it springs upward, stimulated into a bold sapling, quickly placing itself beyond the flames of the next burn. More mature, it lays on bark like blubber, insulating its sensitive cambium from the rush of fire-front heat. It shrugs off even crown scorch, a killing pulse of heat that would decapitate most species. Out of the dead-appearing canopy new buds burst forth from the brown-needled clusters. It's a virtuous cycle: the open-canopied forest promotes wire grass; the wire grass avidly carries fire; the frequent fires flush away competitors to the longleaf. Shutting down fires, however, promotes a vicious cycle. Either the fuels build up to the point of detonation, sweeping all before it, or hardwoods invade and overshadow the longleaf seedlings and eventually overtop the piney woods altogether; or pocosins creep outwards; or the grasses stagnate and shrubs thicken. The longleaf falters. Unburned, it chokes on the stockpiling woody offal, like a stream behind a debris dam.

Early colonists recognized readily the affiliation between fire and longleaf. They picked up the torch from the American Indian and kept burning, partly for hunting and habitat, partly for pastoralism, partly to keep wildfires away, partly just because there was always fire on the land and fire in longleaf took considerable effort to expel. Turpentine operators raked around their tapped trees, then burned the intervening patches during cool winters. Settlers kept bare ground around cabins, and burned off the surrounding rough. Yet the unthinkable finally happened: the fires faded. The trees were logged, the turpentine woods were worked out and felled, the sweep of fires was broken by roads and patterns of settlement (and unsettlement), and as forestry strengthened its grip on the Southeast, boosting pine plantations, by aggressive programs of fire prevention and suppression. Old-growth longleaf disappeared, and longleaf seedlings failed to replace them. The range of the longleaf pine shrank to something under 3 percent of its Precolumbian domain. Its close-coupled fire regime vanished with it.

The near extinction of the longleaf, like that of the buffalo, did not pass unnoticed. Regional observers, foresters like H. H. Chapman at Yale, researchers with the U.S. Forest Service, wildlife biologists—all argued, some belligerently, some in whispers, for the value of burning amid the southern pines, both to contain wildfire and to enhance reproduction; and they made the longleaf the poster child of that campaign. By the early 1940s the national forests of the Southeast accepted controlled burning for fuel reduction. In 1946 W. G. Wallenberg published his magisterial study of the longleaf pine, indelibly fixing its symbiosis with fire. Many observers elsewhere, however, dismissed the spectacle as Southern exceptionalism. Not for another 20 years did the general public sense the deeper arguments for reintroducing fire; not until, that is, passage of the Wilderness Act and the bursting onto the scene of the charismatic California sequoia, with its hoary and prophetic testimonials to fire's ecology.

The Carolina Sandhills NWR needed no such revelations. The refuge exists to preserve the ancient habitat of the longleaf pine. It can only do that if it burns. Refuge managers need to burn, much as cotton farmers in Arizona need to irrigate.

———————

The day is hotter, the air drier. Flames flicker up from ping-pong balls, combusting violently amid tussocks and leaves. Roughly 30 seconds earlier they had resided in a Premo Mark III, bolted onto a Bell 206 helicopter. Then the Premo injected separately into them ethylene glycol and potassium permanganate. Immediately, the balls plummeted to earth, the mix erupted into flame. The oak grove on block 21/01, adjacent to Lake 12, is now sprinkled with flames, like fiery mushrooms sprouting in a time-lapse film. The fires burn hot, merge, and here and there soar through the low canopy. The young oak buds tipping the branches melt like wax.

The scrub oak—turkey oak, blackjack, bluejack, sand post—accepts fire as it does water. Its tough bark sloughs off light fires. Its rhizomous roots resprout from the insulated soil if the surface branches burn away. Once established, it crowds out competitors in the impauperate soils, and then it changes the character of fire's regime, which can skip lightly through its shadowed leaves. This morning's burn, however, is a fire not intended to maintain the oak but to obliterate it. It's designed as a killing fire.

The more singular the oak woods, the harder it is to kindle such a fire. A routine burn simply wipes the floor clean of litter, it crowds out competitors in the starved soils, and then it changes the character of fire's regimen, ready for another leaf fall, another seasonal wash of nutrients. The purging fire appears when the tree has begun to bud, when its twigs are fat with sap, when a hot flame can fry the fresh growth and blister the cambium. It takes several such burns, and they are hard to create because the timing of fuels and growth is out of sync. Sometimes the cycle requires a blast of granular herbicide like Velpar ULW to dampen the vitality of the oak clump and boost the longleaf to where their needles shower the oak litter with sufficient combustibles, like a starter fluid added to charcoal briquets. Granted enough fire rightly applied, the oak dies on its feet, its dry trunks gradually crusting to charcoal as successive fires sweep through. Some 6,000 acres of refuge longleaf are interleaved with scrub oak. And while the extirpation of oak is not the point—oak isn't a toxin, it provides some useful mast and cover for wildlife—it needs vigorous pruning. With the right mix of devouring fire, the longleaf and wire grass can reclaim this site, and others.

Not much lives in the denser oak thickets: they tangle and tear like a briar patch. Squirrels feed on the acorns. A few birds nest. The prime creatures of the reserve, however, thrive amid longleaf. Still, there is a substantial patch of 60 acres, ripe with aging oak, that an early refuge manager believed "natural," since it lay largely outside the longleaf fire regime, and the heavy anthropogenic burning that sustained it. The current staff scoffs at the grove, mockingly referring to its monadnock-like bulge over grassy fields as "the natural area." It is anything but.

The crew spreads out, handling their torches like a turpentiner's axe, searching unit 14/03 for the right niche to kindle. They apply strings of flame along the sloping sides down to a woody thicket as dense as drapery. It is a delicate business: they want to burn off as much of that opaque edge as possible without provoking the whole patch, an intertwined mass of several acres, into a riot of uncontrolled flame. The ideal circumstance is this, the surrounding cane and tussocks are primed to burn while the green woody clump can take flame only along its edges. Such an arrangement allows the crew to force a wall of flame against the neck of the pocosin. If set properly, the flames roar and chip at the green wall, like a sandblaster peeling away stucco. If set poorly, the flames snuff out, or else gush into a blaze like an oil well, spewing a dark venomous smoke.

The pocosin is an upland swamp, an overgrown tangle of hardwoods, cane, vines, and thorn shrubs, a vast woody muck that spreads like green gangrene. Infection by pocosin begins with trapped water, in a low sumpy place. The wetness spares it from fire. The site aggregates organic stuff, a biotic hoarder, gradually spreading up the slopes as its stored biotic ballast allows for more retained water, more hulking vegetation, more propagation by thick tendrils creeping up drainages. The pocosin is the Piedmont forest gone surreal, a dark fantasy, Darwin's "tangled bank" rewritten into southern gothic. For the refuge staff its significance is twofold. Wetland patches hold some rare species—the yellow pitcher plant, the pine barrens tree frog. But they also compete for upland habitat: the more pocosin, the less longleaf. Presently, one acre in nine on the refuge is pocosin.

The topographic texture of the Piedmont is geologically muted but ecologically brazen. The ridges are dry, the ravines wet. The ridges become overrun with oak, planted invasives, weeds; the ravines, with pocosin. The solution to both is to burn them back, with prescribed fire as a fiery weed whacker. Fire's favored tree and the longleaf's disturbance of choice— each will promote a world in which the other can flourish. Since the ridges dry sooner, they are fired first. The pocosins follow. The blackened fingers of ridge hold those flames like a loose fist.

But the pocosin is at least an indigenous blight. The slash plantations stocked in the 1950s and early 1960s are not. At the time of their establishment, the refuge was jointly managed with a belt of state forests of equal area, both overseen by the South Carolina Forestry Commission. When purchased by the Resettlement Administration in 1939, the land had been a denuded shambles. The simple solution, it seemed, was to reforest. That had the additional benefit that the harvested timber would pay for the commission; its felled woods, in brief, had to earn the refuge its keep. The commission planted slash pine, which was here outside its biogeographic provenance. The state forests bristled with regimented columns of alien pine, drilled into deeply scarified land, the latest in the Piedmont's pedigree of soil-wasting row crops. The refuge acquired over 4,000 such acres, roughly what it held of pocosin. The plantations were absurdly, recklessly dense. With 800–1,000 trees per acre, they were, ironically, a fire trap unless vigorously pruned and managed with intensive silviculture.

Increasingly this arrangement jarred with the putative character of the refuge—its political no less than its ecological intentions. In 1984 the solicitor general ruled that the joint tenancy agreement was illegal; eight years later, the lease agreement that bound the Sandhills State Forest with the federal reserve terminated. The South Carolina Forestry Commission retained full responsibility for the state forest; the U.S. Fish and Wildlife Service, for the national refuge. Within the refuge slash pine went the way of other unwelcome exotics, simplified by the fact that it could be logged off and replanted with longleaf, and then burned to help purge the land of residual ecological contamination. Stunningly, despite the extravagant soil wastage of the plantations, despite a biotic flogging that left its skin in bleeding tatters, the land rebounded. This rehabilitation persists. The dominion of fire and, with it, of longleaf continues to spread, like foxes following the dispersion of rabbits to new warrens.

———————

A driptorch passes gingerly around a majestic longleaf with a rectangle of red paint splashed across its trunk. The day is not ideal: the prescription skates close to the limits of toleration. It would be better for burning unit 01/08, flanked by the main touring road, if the wind were a bit stiffer, if the tussocks were a smidgen drier, if the upper atmosphere promised better venting. The smoke curls lazily, too sluggish to punch upward beyond the canopy; and there is a substantial chunk of land under fire, which is to say, overlain by a thick, cloying smoke. The burning must shut down soon to satisfy state-mandated air-quality guidelines. But there are a shrinking number of days available for burning, and the refuge will seize—must seize—every one, and squeeze as much fire as it can from each.

The reason for such tenacity is that the white-painted bands on the trunk mark the site of a longleaf pine inhabited by the red-cockaded woodpecker, the only woodpecker to chisel its nesting cavity in a living tree, and then only if the longleaf is thriving. Why a living tree? Because, in such a fire-frequented place, no dead tree stands for long. Why the longleaf? Because it is the only tree that can thrive amid such a regime. It yields, for the woodpecker, the added bonus that it gushes sap readily (why it was attractive for turpentining), which coats the nest cavity with a slick surface tricky for rat snakes, their chief predator, to negotiate. Why this curious fire ecology triangle? Because that is simply how evolution balanced fire, longleaf, wire grass, and woodpecker. That the bird boasts a signatory red patch on its head aptly symbolizes its fealty to fire. The nesting trees now sport a similar cockade.

There are plenty of other creatures on the refuge—egrets, anhingas, carnivorous pitcher plants, white-tailed deer, glider squirrels, turkeys, otters, bobcats, foxes, the whole menagerie of a landscape once fabled for its fur and game. But the woodpecker trumps them all because alone of that ark it is federally listed as endangered. Whatever the red-cockaded woodpecker needs, it gets. Refuge biologists do what they can to pump up the population: they drill holes for nests, they stock longleaf, they burn. Those flames prevent biotic sclerosis by clearing out the land's ecological arteries. Without artificial cavities and planted pine, the birds could survive. Without fire, they could not.

Those indispensable fires, however, demand as much fussing as the birds; they are themselves threatened with a shrinking habitat. The restrictions surrounding burning are immense; they are becoming nearly impenetrable. There are a limited number of days in which the burning conditions are such that the staff can fire off a block within a daily burning period. Some sites are too wet, and won't kindle or, if they ignite, they spread feebly. Some are too dry, and burn too ferociously to control. Some days the wind is too strong, and fires flee their containment breaks; for some, the wind is too weak, and the canopy traps the simmering heat and the crown gets scorched. Some years the storms come regularly, allowing for a rhythm of wetting, drying, and burning. Some years are droughty. Some are relentlessly wet. Some are prone to hurricanes and ice storms. These factors pertain only to the physics of fire behavior. The biology of burning adds another cluster of conditions. For oak, the fire should be timed with early budding. For pocosin, the burn should happen when the surrounding land can carry a hot fire and the pocosin only singe and abrade away or have its stringy necks burned off. The argument has advanced, too, that the "natural" season for fire is summer, the prime growing season, and that prescribed burns should adjust accordingly. (The refuge believes that fire intensity, not calendar season, is what matters, that the right intensity and the right timing stimulate better germination. Moreover, as opportunities for burning continue to implode, limiting fires to a predetermined season would effectively shrivel the actual fire that gets on the land. Better that some fire happens regardless of season than none at all. They are content to let nature sort out the actual timing.) Then there are social considerations. South Carolina boasts some of the most stringent air-quality regulations in the country—smoke from wire grass or exhaust from automobiles, it is all the same. Only so much can burn at one time, according to the capacity of the airshed to dilute and disperse it. And of course one must control the blaze. Fires must be large enough to introduce economies of scale, yet contained enough not to spill beyond their inscribed borders, and especially not to escape the refuge boundary (which is disked into 16-foot firebreaks). Late-season burns and crown-scorching burns may lead to reburns as fresh needles cascade down to rest on smoldering stumps and there rekindle. Nature's response to one fire may lead, within days, to others, indifferent to any sanctioned prescriptions.

These are formidable factors—and there are others. In most of the United States, ever-lengthening prescriptions are sufficient to stall most efforts to reintroduce fire. Particularly where fire is promulgated as "a tool" for fuel reduction, the public is chary, after having witnessed some spectacular failures, as fires set to eliminate heavy fuels instead fed on them and were stoked into conflagrations. In the Sandhills State Forest outside the reserve, burning has retreated in recent years because the pine straw that would help carry fire is worth more as landscaping décor than even the pine pulp for which the forest was established. Carolina Sandhills NWR reckons it has fewer than 30 days a year suitable for prescribed burning, these scattered from mid-January to late April; and this number is dwindling. Yet the refuge manages to fire the land. One reason is a skilled staff. The deeper reason is the red-cockaded woodpecker. What the woodpecker needs, it gets. The woodpecker and its preferred nesting tree, the longleaf, demand a lot of fire.

The last rain fell three days ago, a steady, drenching downpour that soaked the woods and piled up on asphalted roads. It seems incredible that the land could hold flame so soon afterwards, yet it does. The torches put down flame on unit 21/02 in long streamers that snap crisply through the stalks and tussocks fringing the Visitor's Drive. Far from being extraordinary, this regimen is routine. The refuge's preferred burning season coincides with the rhythm of winter storms. In mid-January and February cold fronts pass through with some regularity; the land dries, the winds calm, the refuge staff burns; then another front douses the smoldering debris, the winds bluster, and the cycle begins anew. When the tempo of rain becomes more spasmodic as storm tracks shuffle northward, the occasions for burning become more erratic.

Again, biological factors apply as well as climatic. A burn too early, and the woods have not yet lapsed into dormancy and shed their leaves and quota of needles. A burn too late, and the green up overwhelms the dead litter and stalks with a flush of moisture that makes fire difficult. Still later, amid the heat of summer, particularly with a dry spell or outright drought, and the flames might billow uncontrollably—incinerating what they should only scorch or bolting out of the refuge altogether. Summer, too, is the season for western wildfires, and the refuge fire crew often heads west to help staff those firelines. These circumstances apply widely throughout the Southeast. What the Carolina Sandhills have that makes this pattern of burning possible is its eponymous sand. The refuge has a lot of it.

The sand's thickness varies from 2 to 40 feet, the geologic residue of an ancient coastal plain into which rivers debouched their sediment. It is, in a sense, an interior, ancient analogue of Carolina's coastal barrier islands. The sand drains quickly, and shallow-rooted plants, those that carry the flames, find that the land leaches even heavy showers briskly. The storms fill the soil with water, the sand whisks it away, the fires feed on the fast-drying growth. Without the sand the cadence of wetting and drying would be far more fickle. The burning season would become atonal rather than melodic. Sand and storm thus set the fundamental cadence for fire.

Such a surface of friable sand is unstable. What holds it together is the dense weave of roots, the equivalent of dune grass. Without that web, the dry sand acts, as the locals put it, like a "bowl of sugar." But as with prairie sod, agriculture ripped away the protective cover. Clearing and plows peeled off and uprooted that tough husk, and set into motion a wave of erosion that sent hefty chunks of the sandhills to the sea. Row crops— cotton, corn, tobacco, pine—and poor plowing techniques worsened the excavation. By the 1930s the land was derelict; its farmers impoverished, its biota a ruin. Its ecological pieces resembled the jumbled letters in a Scrabble game box. People and creatures had become mutually exhausted sharecroppers. In 1939 the Resettlement Administration bought out and moved the farmers off 90,000 acres, dividing the allotment between state forests and a national wildlife refuge. The reconstruction of the land commenced.

What kept the geologic dunes from total disintegration were hardy ecological survivors, the longleaf where it clung tenaciously to its old habitats, and especially the scrub oak, as grasping as a Snopes. Folk wisdom held that only they lashed the sands together. Remove those roots and the land would collapse, like stacked hay with its baling wire pulled out. Any scheme of rehabilitation had to replace, not simply expel, the scruffy oak with the more suitable pine. The most pliable medium for this exchange was fire. Paradoxically, the sand that encouraged the decomposition of the land also promoted its reconstitution because its poor nutrients favored tough, fire-hardy species, its hydrology kept them dry, and its human residents kept them in flame.

———————————

The flames are everywhere, bursting over unit 09/02 and its longleaf-oak cluster like April dogwood into blossoms. Crews on the ground lay down strips of fire along roads and around sensitive patches. But the serious kindling, like a benign barrage, descends from the helo circling overhead. With so few truly great days for burning, the refuge needs to magnify the punch of every one that happens. Today is a marvelous day to burn. They cluster—Mike, Mark, Clay, Dave—along with their ATVs and the Bell 206, like bees to fresh blooms.

Burning remains more art than science, but before it was either, it was a practice, or more specifically, a rural craft like hunting with dogs or harnessing mules. It's an ancient human skill—among those that most define us as creatures. Rural burning died out, however, when the rural economies that sustained them became marginalized, and then replaced outright; when mills and factories supplanted sharecropping, when ruralites moved to cities, when exurbanites began to recolonize suburban fringes, when hunting and fishing became a recreational pastime, not a necessity, when golf evolved into a middle-class sport and challenged upper-caste hunting plantations, when racing stockcars replaced running hounds. Rural burning persisted far longer in the South than elsewhere in the United States, just as its rural landscapes clung longer to the old ways. Today roughly 80 percent of controlled burning in the United States resides in the Southeast. The South, not the telegenic West, commands the lion's share of American fire. Carolina Sandhills NWR cycles within that larger orbit.

In the early years, the refuge and surrounding forests confronted an out-break of wildfire. Uncontrolled flame was as much a part of the wrecked landscape as weeds and eroded gulleys. The war years worsened the scene because there were few personnel to fight the flames and, worse, the reserve became the site for military exercises, particularly for artillery and aerial bombing. The fire load shot up. In one year alone, more than a fifth of the refuge burned. In 1947 formal fire protection came to Chesterfield County; fire laws, previously ignored, became enforced; the refuge and Forestry Commission urged fire prevention to the public and fought fires. Burned area on the refuge dropped dramatically. Even during record outbreaks of fire throughout the state, the refuge passed unscathed.

But as wildfire receded, the refuge sought to replace it with controlled burns. This was the fire component of reestablishing longleaf. In the early 1960s more acres were control-burned than burned by wildfire. Annual reports celebrated controlled burning for "habitat improvement" for "deer, turkey, and other species, while greatly reducing wildfires." Then a massive ice storm in 1969 shut down the program for nearly five years while the refuge and state forests wrestled with horrendous woody debris strewn over the land like straw. In 1970 the red-cockaded woodpecker was declared an endangered species, and three years later received protection under the Endangered Species Act. The burning resumed, reaching a plateau of roughly 1,500 acres a year. In 1975 plans called for burning uplands on a five-year rotation. This was still a far cry from what the longleaf demanded. The breakthrough came in 1982.

Enough burning had tweaked enough prime longleaf that the refuge could rely on heading fires and spot ignitions rather than fire-"plow" a site with furrowed strips of backing fires. It acquired access to a helicopter, equipped with a helitorch that dribbled gelled gasoline, which meant it could hit those sites with fire when the opportunities occurred and it could enlarge burning blocks to natural barriers rather than tractor-plowed firebreaks. It could do more with fewer people and plows. By then, the red-cockaded woodpecker was endowed with an official recovery plan, which tied its survival to that of the longleaf pine. Nationally, the Fish and Wildlife Service had been stung the year before by a double-fatality fire on the Merritt Island refuge in Florida; funds materialized to train a fire-specific staff for the refuge network.

The refuge now had motive, means, and opportunity. Burning escalated. Aerial ignition became, as it were, a "force multiplier." Another review in 1992 urged a three-year rotation for burning. Annual burned area bumped up to 5,000 acres. In 1996 the refuge hired its first dedicated fire management officer and support staff. Two years later it claimed exclusive use over the helo and Premo ignition systems. By 2000, annual burning leveled off at an average of 15,000 acres, a third of the refuge. Staff began to push for a two-year cycle on prime sites and to vary the sequencing of burns on particular sites, allowing, for example, a late-season burn to mix with dormant-season burns. Still, some years offer more occasions than others. In 2003 a cadence of late-season rains allowed the refuge to burn off backlogged sites to the tune of 20,000 acres, almost 45 percent of the whole.

Its staff made it happen: it arranged those scattered Scrabble pieces into words, and the words into sentences. The refuge fire staff knew the refuge, and they knew burning. They knew how to seize the moments that place and weather presented. They burned the way a white-tailed deer browses; constantly and selectively; moving, pausing, returning. When the refuge proper couldn't burn, they burned throughout the region and the refuge system. When conditions were poor at the Carolina Sandhills but good at Fort Jackson Military Reservation outside Columbia, they trekked to Fort Jackson and burned. When conditions aligned at the Pee Dee NWR, they burned at the Pee Dee; at Piedmont NWR; on Florida refuges; on the nearby state forests; wherever. During the summer they headed West to work on the big fires in California, Montana, Arizona, or beyond. They had become as migratory as cranes, searching out flames along the seasonal flyways of America's fire geography. Not calendars, not bureaucratic prescriptions, but those fleeting occasions where flame and fuel could meet, a biology of fire as sensitive as the ripening of pine barrens gentians. They were fire foragers, as dependent on fire as longleaf and red-cockaded woodpeckers, but with this vital difference: they could start the fires they required. As long as their peculiar bureaucratic habitat remained, they could thrive, and fire would flourish.

―――――――――

It is barely midday, hot and the sun high, and the burning on unit 21/01 has shut down. It won't be restarted in the afternoon. It won't be restarted the next day. The shutters have slammed closed on prescribed-fire's seasonal windows.

The refuge has met its acreage quota; pushed its smoke allowances to the limit; brushed up, as April ripened, against the end of the controlled-fire season. Likely there are fires to set elsewhere, and probably fires to fight. But the refuge's fire program involves more than burning. It is not simply that fire prepares habitat: it is that fire requires its own habitat in order to do the work it should. Flame is catalyst as well as chisel: it will synthesize, not simply shape, its surroundings. Mangled settings may only yield mangled fires. Staffers may need to ready fuels as surely as they drill out nesting cavities for the red-cockaded woodpecker or weed out invasive pocosin scrub. Tinder without ignition simply rots. Ignition without kindling only sparks. The restorative fires need both in sync; that is the ecological value-added service the refuge's fire staff provides.

The contrast with the western fire scene is striking, particularly the contrast between western yellow pine and southern yellow pine, between ponderosa and longleaf. It is not simply that one place burns with prescribed fire and the other with wildfire. There was plenty of wildfire once on the Carolina Sandhills refuge, and most western landscapes knew millennia of controlled burning by the peoples who inhabited them. But the comparison of recent decades is eerie. At the time when the federal agencies were moving to revise their policies to better promote fire, the refuge succeeded in making that conversion, not by policy pronouncements but by hard practice. Wildfire shrank, controlled burning rose. Repeated firing

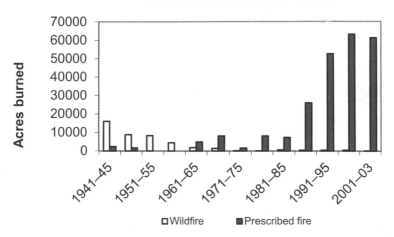

FIGURE 1 The chronicle of burning at Carolina Sandhills National Wildlife Refuge. One of the most dramatic examples in which prescribed fire has substituted for wildfire, while increasing the burning overall.

peeled away the fuels that might otherwise have powered the kind of biota-scouring holocausts that have become commonplace in the West. It isn't enough to appeal to drought as the cause of the West's fire plague. Prior to the winter of 2002–03 the Carolinas had endured a multiyear drought that left most of the Piedmont with a rainfall deficit equal to a full year's precipitation—this imposed on a forest far more densely loaded with combustibles than those typical of the West.

Plot the graph of the refuge's overall burning and you see a decline, leading to a steady state, then a sharp burst in area burned as controlled fire multiplied its presence. Plot the burned area for the western national forests for the same period, however, and the graph is uncannily similar. A bubble of burning during World War II, a leveling, and then around 1970 a rise in area burned not unlike that for the refuge. The difference is that the refuge controlled its flames, and it has reached a kind of ecological plateau. The western national forests were fighting wildfire, and the burning showed wild peaks and valleys, a kind of cybernetic hunting as the system sought, unsuccessfully, to find an equilibrium. Clearly, part of the reason for these violent swings is a buildup of fuels. Clearly, too, the tipping point for both systems came around 1970. The refuge struggled to overcome a forest temporarily trashed by ice. The western forests wrestled with a heritage of fire exclusion. Probably this was the last time reform could have come with relative ease.

This comparative parable is far from the whole story, however, and it misses the really interesting contrast. The refuge has been able to reconstruct itself. The western forests, subject to far less havoc, have not. The extent of landscape tweaking, meddling, fidgeting—felling, planting, plowing, burning, hunting, spraying—on the Carolina Sandhills NWR is staggering. By the 1930s humans had pretty thoroughly deconstructed this landscape, which lay in shards, so much red longleaf scrub wire grass pocosin oak cockaded sandhills. The land, one observer remarked, had been "rode hard and put away wet." By 2000 they had largely rebuilt it; not always wisely, not without having to intervene again to correct blunders like slash pine, not without repeated revisits to sites as knowledge and errors accumulated. They could not have resurrected this land without fire, which supplied the ecological grammar to rewrite its scrambled sentences; but fire, unaided, could not have restored a landscape so shattered. Instead, the refuge would have become infested with wildfire, as it was during the war years. Fire might have made worse what was already bad. Burning a shambles does not leave a gleaming new building; it may leave a smoking shambles. The land held the right kind of fire because it was the right kind of land. Fire and fuel—both fell under the ardent care of the refuge staff.

At Carolina Sandhills, the refuge's task is to showcase fire as it does the woodpecker. One fire staffer states their philosophy simply. It is their job, he says, "to put out the fire"—that is, to place it on the land. Nature will sort out the particulars. It will organize those biotic fragments into phrases, and if all goes well, into sentences that convey a theme.

SLOUCHING TOWARD GATLINBURG

Things fall apart; the centre cannot hold;
  Mere anarchy is loosed upon the world. . . .
. . . .
And what rough beast, its hour come round at last,
  Slouches toward Bethlehem to be born?
        —W. B. YEATS, "THE SECOND COMING"

A fire in the southern Appalachians, a regional entrepôt burned—where did this come from? Is it a freak like the New Madrid earthquake, or the annunciation for the birth of a new abnormal?

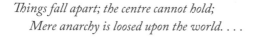 FIRE BEGINS IN WILDLANDS. It burns through a shocked countryside. It blasts into a town that, if old, had not known such burns for a long time and, if new, had never experienced anything so savage. Embers kindle new fires far ahead of the front, flames leap from house to house. Until the winds die, fire protection forces are helpless. A postmortem points to contributing factors that residents believed had been expunged from vernacular landscapes. Now they have returned, like a reemergent plague.

The Chimney Tops 2 fire that rushed out of Great Smoky Mountains National Park and ripped through Gatlinburg, Tennessee, and clawed to the borders of Pigeon Forge on November 28 is novel only in its time and place. Thanks to a tenacious drought, it occurred outside what is regarded as the normal fire season. It ripped over a landscape that has not known fires of this sort for a century. It's the kind of fire that happens somewhere else. Only this year it came to Gatlinburg.

The American fire community once knew burning communities as a rural fire problem, most explosively associated with frontier landclearing. In the post–World War II era, cities and suburbs sprawled across the countryside. The buffer between the wild and the urban shrank until, in places, no buffer existed. The first expression of the new fire was probably the 1961 Bel Air-Brentwood conflagration that burned in Los Angeles. Here exurbs had pushed into brushy hillsides.

In 1986 the problem received an official if geeky name, the wildland-urban interface fire, coined in California. The U.S. Forest Service and the National Fire Protection Association joined in a national campaign to alert the public about the menace and about the measures communities needed to take to protect themselves. Initially, it tended to be defined primarily as a wildland problem because that's where most of the fires came from.

But observers could as easily have picked up the other end of the stick and defined the problem as an urban one, with exurbs built without regard to zoning and houses erected without regard to codes. The vulnerable places were patches of cities with peculiar landscaping. In 1991, America experienced its worst urban conflagration since the 1906 San Francisco earthquake as fire poured through the Oakland Hills across the bay. Here the problem was a thickening of vegetation in open space next to and within a long-established town.

Still, it seemed a California quirk. And then it wasn't. Wildfires began taking out houses in Arizona, Montana, Colorado, Oklahoma, Washington, and in 2011 Texas, on a monster scale—all places where Americans were living, as in California, on an edge between the feral and the tame. But fires were also boiling over in Florida, where developments pushed inland into some of the most fire-thirsty biotas in North America. They began spreading across the coastal plains in the Carolinas and Georgia. In 2007 fires broke out of a droughty Okefenokee National Wildlife Refuge and threatened the surrounding countryside. The bad burns were no longer quarantined on the West Coast or the semiarid, semiempty landscapes of the West. They were even moving out of the coastal plains for the mountains. A widening sprawl was colliding with worsening sparks.

The narrative changed. For decades it was convenient to treat those outbreaks as a freak of western violence, like a grizzly bear attack. The prevailing narrative was one of dumb westerners building houses where the fires were. Now the fires seem to be going where the houses are. The National Association of State Foresters reckons that 79 percent of America's communities at risk lie in the South.

We won't know all the contributing causes to the Gatlinburg blaze until the pyric postmortem is complete. It may be that the Chimney Tops 2 fire was a low-probability, high-cost event that was ever latent yet unlikely. But widen the aperture a bit—think on longer axes of space and time—and the odds seem less extreme.

The southern Appalachians have long known fire. The record from ecology, scarred trees, and history, both written and oral, and for that matter, folklore, testifies to a regular regimen of surface burns. A recent summary of Appalachian fire history concluded that "fires had occurred at short intervals (a few years) for centuries before the fire-exclusion era. Indeed, burning has played an important ecological role for millennia. Fires were especially common and spatially extensive on landscapes with large expanses of oak and pine forest, notably in the Ridge and Valley province and the Blue Ridge Mountains." Then large-scale logging and slash fires swept over the mountains in the late 19th and early 20th centuries, leaving apocalyptic landscapes that became poster children for conservation. The 1911 Weeks Act that allowed the young Forest Service to expand by purchasing new lands had in its crosshairs the wrecked watersheds that followed industrial-strength slash and burn in the southern Appalachians. National forests were acquired patch by patch. The Great Smoky Mountains were set aside as a national park in 1934. That was the year of the last big burn, at 10,000 acres. With the help of the Civilian Conservation Corps (CCC) the park gradually installed a modern infrastructure for fire protection.[1]

Slowly the land recovered. Fires were fought, fires were no longer set. The forests regrew. The new woods enabled "mesophytic plants to expand from fire-sheltered sites onto dry slopes that formerly supported pyrogenic vegetation." The new woods were generally less flammable than the former ones. The woods thickened, deciduous trees no longer shaded the surface in fall and spring, logs piled up in and around streams, compromising their role as fuelbreaks. The memory of expansive burning and big blowups passed. The economy shifted from production to services, from working landscapes to amenity communities. New villages sprouted, old ones were reborn. The rural countryside that had once burned often but had lots of workers to hold fires in check gave way to exurban hamlets, tourist attractions, overgrown forests, and lots of people incapable of helping. If the West was an interface, with sharp boundaries between public and private land, the Southeast was more of an intermix, an omelet of town and country, new and old. Now climate change promises to further juice the prospects for bad fires.[2]

This looks a lot like the western fire scenario. Park officials chose to monitor the fire, which was in a remote site and which typical weather, they assumed, would shortly douse. There seemed no justification for putting firefighters into tricky terrain. It was the kind of strategy developed years ago in western parks, but one that was being replaced by a more proactive box-and-burn approach because blowups happened quicker than monitoring fire organizations could respond, especially for off-season fires. But a deep drought was on, the winds suddenly blew with savage strength, there was little on-the-ground suppression capability. The flames burst through Gatlinburg, and beyond. Fourteen people died, 140 were injured, 14,000 were evacuated in often frightening circumstances, 1,700 structures were destroyed. Officials could be forgiven if they thought this was the kind of fire that happens in California. Exactly. Because that's what it was. It was the kind of explosive burn that has, over the past couple of decades, leaped across the West, and now may have jumped the Mississippi and into the Appalachians.[3]

It resembles California in another respect as well. It has been dismissed as arson, human malevolence perhaps abetted by mistaken management; the blame falls on the teenage boys who set the fire on Chimney Tops. In fact, the county attorney's office declined to prosecute because the eruptive winds toppled down power lines, which also set fires on the outskirts of the city, and it was impossible to assign specific blame. The fundamental problem, that is, was structural. It was how people had chosen to live on the land and how, here, they had forgotten about fire on that land. Gatlinburg might well have burned without Smoky Mountain's handling of its awkward burn.[4]

If unchecked what unfolds in the southern Appalachians, and elsewhere in the region, will follow the early California example before the state learned what had to be done. But there is no reason why the East cannot pick up the good practices instead of the bad. It can begin with what the West has learned over the past 50 years of surprises that should have surprised no one. It can happen here.

Most of the larger environmental forces are aligning in ways that promise to worsen the scene. There is not much we can do immediately about global climate or the national economy; we can only tweak them slowly, nudging them to better outcomes. Meanwhile, we can harden houses, strengthen defensive buffer zones, bolster firefighting capacity, and reinvent a working landscape to hold a new middle ground. We can accept that eastern landscapes that had known fire in historic times will experience them again—will need to burn again. We should adapt fire strategies appropriate to what promises to become a new abnormal. Otherwise the rough beast of the western fire scene will continue to slouch eastward.

FIRE SIDEWAYS

NOT ALL FIRE NEEDS to be deadly serious, or existentially earnest. Sometimes a touch of humor, a frame rotated 90°, or picking up the other end of a stick can yield insight not accessible to the formal and the academic. The gathering of essays that follows is written in that spirit. They disperse rather than cluster. They come at ideas a little out of the expected.

AN ENDANGERED
PROCESS ACT

Prescribed fire and managed wildfire need political leverage. The Endangered Species Act shows the power of legislation that mandates prescriptive, not just proscriptive, action in some places. Fire needs something equivalent if it is to return to more than a pittance of its former splendor.

FREE-BURNING FIRE HAS BEEN on Earth for 420 million years. Hominins have used it since *Homo erectus*, at least for cooking and most probably for cooking landscapes as well. With the advent of *Homo sapiens*, fire found a keystone species to help maintain its role as a keystone process.

So when, a couple of centuries ago, that creature began to derive its firepower by burning fossil landscapes, which meant the substitution of closed combustion for open kinds, the switch became a swerve that has rippled widely through the Earth system. The additional combustion, or more precisely its effluents, has destabilized the climate. The removed combustion has helped unravel many biotas that had evolved to accommodate particular fire regimes. Fire has retreated into reserves, not unlike so many endangered species.

But fire is not a species, and there is no legislation to mandate practices that assist it where it is threatened or endangered. In fact, the Endangered Species Act helps to account for the regional distinctions and ambivalent outcomes in controlled burning. The firebirds of the eastern United States (think red-cockaded woodpecker, Florida scrub jay, Bachman's sparrow) all work to boost burning. The threatened and endangered species of the western United States (sage grouse, spotted owls, marbled murrelet) all push toward suppression. There is no legislative counterforce to argue for prescribed burning.

America's great cultural revolution on fire, which boiled over during the 1960s and became encoded into agency manuals during the 1970s, allowed for the reintroduction of fire. It left the choice to burn or not with local park superintendents, forest rangers, or refuge managers. Some seized on the opportunities; most ignored them. The record clearly favors those who avoided mistakes rather than those who boldly tried and failed. The choice to burn or extinguish yet remains with agencies and administrators. Suppression persists as the default position. What the revolution lacked was an explicit mandate to restore fire. It was as though the parallel civil-rights revolution had encouraged racial integration, but did not require it, in which case the bad old ways would find new means to continue.

If fire truly matters, if fire's presence is integral to ecological integrity of many biotas, if open flame is a more benign medium for humanity's firepower than internal combustion, if, as formal research and long familiarity argue, fire is a keystone process, then we need legislation or policy that stipulates that it must be maintained or reinstated—not that fire's restoration to historic levels would be nice, but that it is necessary, and the failure to restore is illegal. Such legislation lies behind Parks Canada's successful fire project. Its absence helps explain why America's national parks continue to stumble and fall behind.

We need, in brief, an endangered process act. Until fire restoration—or its successor project, fire resilience—has the force of law behind it, we will continue to choose the path of least resistance, which is also the path of least resilience. The only fire allowed will be wildfire. Sometimes individual will is not enough. Sometimes we need to voluntarily subject ourselves to a collective compulsion to do what we need to do.

SUBPRIME FIRES

A Socratic Tutorial

Thoughts provoked by the unholy alliance of megafires and financial meltdowns, in a less-than-Platonic dialogue.

OVER THE PAST 40 YEARS California has experienced a series of what nonresidents regard as earthshaking wildfires, at least a 6 or 7 on a Richter conflagration scale. They have by now become a part of the environmental background, like kona storms, tremors, and Santa Ana winds. I wanted to know how Californians were coping, so I asked my oracle on all matters Californian, Harvey. Harvey had spent a few seasons with the Loco Lobo Hotshots before trading in his brush hook for a realtor license. Hotshot and hustler, it was a perfect California perspective.

Harve, I asked. How are Californians coping?

"Quite well," he replied. "The people know the fundamentals are sound. They have learned to manage—in fact, master—fire risk just as they have economic risk."

When I looked puzzled, he continued.

"Fire and financing—they're mirror images. It's easy to go from one to the other."

When my brow still furrowed, he sighed. He knew this would take some background explanation, but then we go back a long way. He gathered his thoughts. "It works this way. In the bad old days, there was a horrific fire problem along the frontier. Farmers and ranchers hacked and burned their way across the continent. Mixing outbuildings, towns, and woods was, to say the least, combustible. A place would bloom one day and burn the next. It was okay so long as the frontier was big. A hamlet here, a village there—a few flames were the price of a red-hot realty market. No one wanted to stop the fires because that might slow down the frontier deal making. It was a land rush. That's how the country got developed. That's what made us what we are."

He paused. I could sense a change coming. When he began again, it was Harvey the hotshot speaking.

"But then the frontier slowed, and people became alarmed over a timber famine. If these fires kept going, they would burn down places we wanted as permanent forests. They would burn down giant sequoias and Yosemite Valley. So gradually lines were drawn and wildfires were fought and pretty soon flames no longer stampeded into the parks and forests. It took a long time. There was a huge crash and enormous big fires into the early 1930s. But Roosevelt's New Deal used the Civilian Conservation Corps and emergency conservation money and a lot of planning and segregated the two landscapes. Each had its own kind of fire, and both became pretty good at keeping the flames in bounds. They did with land what they did with the economy, keeping the bankers out of the stock market."

He paused, and then began again. This time it was Harvey the hustler speaking. "Well, people were bored living in designated places. They wanted a new frontier. They wanted to start it up again and there was a rural landscape that wasn't really doing its bit in the market. Of course we're a lot smarter than the fogies in the Old West and we have a lot of 'new instruments,' so the quick studies decided they would tear down those barriers, and open up those landscapes, and give people what they wanted—a little house in the big woods. (Actually, a lot of people wanted a big house anywhere, and the feds were happy to give them a tax write-off to do it. It was a real win-win deal.)

"So the barriers came down, slowly, then rapidly, and houses spread like cheatgrass. A real boom for the economy. Couldn't write subprime mortgages fast enough."

Yes, I said, but didn't those exurbs spread into subprime landscapes? "Isn't there a danger from fire?"

"Some," he said. "But we have learned to spread the risk. It's the same as securitizing. We chop the landscape up into finer units and insert more and more houses. Even if a thousand houses burn in a year, that's nothing compared to a million new constructions. The odds of any one house burning are slight."

I might not agree if it was my house, I pointed out.

"Yeah. I see your point. But that's because you don't understand the way fires work now. It's different than in the old days. Today we aren't clearing land with fire. Those exurbs aren't starting fires. The fires, if they come, break out from the public lands. The feds have to fight them. They have to bail us out.

"And if a fire does leave those protected lands, there are all these inter-locking agreements with the states, and counties, and cities, and anyone else that allows the fight to continue to your doorstep. That's real security. It's like globalizing the monetary system. The risk is divided into fine parts, responsibility is diffused everywhere, and the official response to an outbreak is unlimited. FEMA will step in. It's like the Greenspan Fed flooding every Wall Street hiccup with more money. The market just roars back bigger than ever."

He was going now, cutting his new line as quickly as he did in the old days. "If I were a homeowner, or second-home owner, and there was a big wildfire bearing down on my place, I wouldn't worry. I know they would stop it. They won't let it happen, any more than they would let Wall Street burn down. Money is no object. They spend everything, and then get more. And if it does burn, you'll get help to rebuild. No one wants the boom to end.

"It's not just California," he said, with touch of huffiness. "Hell, hurricanes mean nothing to Florida and Louisiana. No serious economist would argue we shouldn't build in floodplains because even if it floods, the flow of new money to rebuild will jack up the GDP. It's the same with fire.

"Of course there are always a few whiners and nervous Nellies. They want those barriers back. They just don't get it. We'll never have another Peshtigo or Hinckley burn. We have firefighting forces now they never dreamed of in the 1870s and 1890s. Those kind of catastrophes will never happen again, any more than we'll have another stock market crash like 1929."

You must be right, I said. The authorities wouldn't let all this happen if they didn't know they could prevent a genuine meltdown. That's what they did in the Great Recession. I guess I just fret about that disclaimer in the fine print.

"Don't worry," he said, soothingly. "The lawyers make us put that in. It doesn't mean anything. Houses can go up, and houses can burn down. What does that mean?"

SPOTTING PACK RATS

Remembering that biology lies behind the physics of fire behavior with some eccentric ecology as illustration.

F EW TOPICS IN FIRE BEHAVIOR are as complex as spotting, and few illustrate better the promise of fundamental science. Almost all the basics of wildland combustion come into play. To model spotting requires an understanding of fire intensity, which creates firebrands; of torching and crowning (and perhaps whirling) to loft them; of wind flow to carry them; of smoldering rates that keep them alight; and of ignition probabilities that determine if they can kindle in fresh fuels when they land.

None of these phenomena are unique to spotting, which just arranges them in particular ways. But understand spotting, and you understand most of fire behavior, since fires recombine the same processes in various ways. So while spotting may carry particles of fire beyond the flaming front, the science of spotting does not leap outside the realm of fire's fundamentals or the relationship between research and practice. In this way, spotting can serve as a model system not only for fire behavior but for the value of research into fire fundamentals.

In the chaparral of the San Gabriel Mountains, however, the logic of spotting obeys a somewhat different dynamic. Those on the ground— the Chilao Hotshots, for example—reckoned that spotting was often a byproduct of pack rats. Their nests were a shambles of kindling that readily received embers, ignited, burned hot, sparked escapes, and flung out new brands. Sometimes even the pack rats themselves caught fire and scampered across the hillsides scattering fire in an echo of many Native American legends. The geography of spotting followed the distribution of nests. The older the chaparral, the denser the nest sites, and hence the more likely the landscape to burn, since spotting makes control notoriously difficult. If the pack rats were obliterated by a plague, spot fires would shrivel away.[1]

Of course the fuel moisture content of nests matters, and if there is no wind, firebrands cannot travel far, and the intensity of any combusting nest naturally reflects the chemistry and particle size of the twigs that tangle into a nest. So far as physical modeling goes, the nests are simply caches of combustibles. But the deeper reality—the real fundamentals—is that those fuels are biomass, that they burn with a biochemistry, and that the arrangement of fuel strata reflects the evolutionary ecology of that biota, and in this case the nesting habits of pack rats. Without pack rats, spotting would look different. The behavior of pack rats thus influences the behavior of free-burning fire. (For the flip side of this observation, point to burrowing marsupials in Western Australia who so disturb the surface that they break up the continuity of litter fuels. The wholesale destruction of the creatures has boosted fire spread. For an American analogue, consider prairie dogs.)[2]

Or to put it differently, the physics does not underlie the biology of the scene. The biology underlies the physics. And both do not derive from first principles but are historically constructed: their boundary conditions and actual arrangements are the outcome of past events that are idiographic. They cannot follow directly from first principles.

───────────

The pack rats are not alone. In the Transvaal of South Africa the classic explanation for the patchy geography of woods and grassland is the presence of free-burning fire powered by fuel buildup, drought, and wind. Routine fires sweep the savannas free of seedlings; hot fires, fueled by older woodlands fluffed with unburned fuelbeds and leached of moisture by drought, drive back the tree line. In both instances the resulting patchiness of woodlands derives from the physical properties of fire. What explains fire behavior equally explains fire ecology.

Yet a study in the Nylsvley Nature Reserve discovered a different dynamic for the trees, both in their establishment and in their survival. Neither of the major species (by dominance, *Terminalia sericea* and *Burkea africana*) takes root in termite mounds, which leaves them in the more pauperate soils typical of frequently burned biota. The trees survive the light washings of flame easily. There appears, nonetheless, a tendency, after a canopy matures, toward more shade-tolerant species and a dampening of fire properties. If the same processes, including fire, continue, the woods ripen into denser stands of *Dombeya rotundifolia*. That the savanna (and lighter woods) survives means that some additional factor has to intervene.[3]

The explanation offered by a physical model is that fuels build up and, during dry spells, that denser stash of combustibles feed fires that overpower and flush out the developing woods. But once mature, which is exactly when the model would predict maximum fire intensity, when physical factors would presumably overwhelm any biological capacity to resist, the trees nonetheless survive. Heat alone seems insufficient to kill them off. Something else must operate to perpetuate the patchiness.

Elephants do nicely. But elephants have long vanished from the Nyl Valley. What serves instead are porcupines. The injuries they inflict may kill trees outright by girdling, but it is enough that they create small cavities, lacerations, and ring-barked scars that fill with pitch which then harden into a daub of highly flammable fuel. Unscarred trees survive; injured trees burn. Without that leverage, the flames would crash into and wash around toughened bark with no lasting effect.

In brief, fire can fell the big trees because they have been readied by porcupines, which conveniently prefer the late-succession *D. rotundifolia*, a light-barked tree easily killed by girdling heat. Porcupines also attack *B. africana*, but they are tougher, and require an average of four pitch-powered fires before they succumb. That the porcupines preferentially gnaw at the end of the dry season (when *D. rotundifolia* begins an early sap flow) aligns fresh pitch with the annual rush of flame.

A model that would reduce the landscape to fuel might identify the scarred woodlands as potential caches of combustibles conveniently daubed with starter fluid. But it would miss the real dynamic of this woodland savanna, which is what creates the fuels and microclimates within which fire must burn. The processes that join fuel to flame are the termite and the porcupine. The one determines which trees, and where, take root, and the other, how long they survive. One could then add *Homo sapiens* as a primary source of ignition. The sculpting fires would seem an emergent property of the biota, not simply of climate, terrain, and chunks of inert hydrocarbon.

The interaction of savanna and woods is not the outcome of heat from a flaming front slashing into woodlands but of a more nuanced collusion between various organisms with fire as a medium of exchange. The physics can only explain the outcome once the ecology has done its work. The fiery felling (and management) of *Burkea africana* woodlands is not about managing fuel so much as managing porcupines.

━━━━━━━━

On the 2011 Wallow fire, the largest in Arizona's recorded history, did particular damage to old oaks. This surprised many observers, since the fire community generally regarded oaks as irrelevant in the fuel composition of the landscape. They were deciduous, their canopies never burned, their fallen leaves made flimsy fuel. It was the ponderosa pine, their surfaces piled with needle-cast and windfall and with ladder-fuel understory, that was the fuel problem. The Wallow fire, however, burned, as wind-driven spring fires often do, very shallowly except where mixed conifers clothed ridges and hillsides exposed to the full rush of the wind, in which case the wind blew the flames through the crowns. This was hardly possible with mature oaks.

The reason for old-oak burning is that over their lifetimes the oaks acquired cavities. Some were basal catfaces, the scars from old surface burns. Most resulted from broken branches. Those cavities became prime sites for owls and especially Arizona gray squirrels. The squirrels filled the cavities with cone middens while outside they gnawed pinecones whose droppings gathered at the base of the trunk. That pile of scales, like pack rat nests, became the site of ignition, stubborn burning, and eventually the death of otherwise healthy oaks. (What the fire didn't take, firefighters did by felling the smoldering trees.)[4]

The coefficient that reconciled fire behavior with fire severity was the Arizona gray squirrel. It's hard to derive the impact of squirrel middens from the first principles of fire physics.

When a fire is blowin'-'n'-goin', the physical model of fire behavior is the only game in town. But management on a landscape scale does not follow from the formulas of fire behavior. Rather, fire behavior derives from the properties of the landscape, and many of the critical factors are biological.

This observation is so commonsensical and universal that it would seem axiomatic. For a variety of reasons, however, fire science too often wishes to reduce its complexities to physics alone, or at least to those properties of physical geography that underwrite fire behavior. The assumption is, Know those geophysical parameters and the principles that govern them, and you can derive fire's presence on Earth. Moreover, of all those physical factors, climate is reckoned supreme. By setting patterns of wetting and drying, climate determines primary productivity; by imposing bouts of deluge and drought, it transforms that biomass into fuel; with lightning, it starts fires; with wind and synoptic weather it drives landscape burning. The fire regime is an epiphenomenon of physical geography and the principles of fire behavior. Fire ecology is so much flotsam and jetsam floating on the deep swells of climate.

This is nonsense. The most dramatic changes of recent times involve the sudden removal and introduction of species. The ejection of elephants and rhinos from the African savanna, the eruption of cattle and sheep in the American West, the spread of invasive grasses—these have restructured fire regimes wholesale without a change in climate. They occur far more rapidly and with actual specific outcomes than anything hypothecated by global warming. And we won't even go into the causes of global warming, which trace back not to some cosmological process but to the actions of a single species, *Homo (putatively) sapiens*.

There are hundreds of studies that examine fire's effects on small mammals. The assumption throughout is that fire burns and creatures react. The reverse, however, is equally true. Small critters live, and fire responds.

The physical model is not wrong in some ontological sense. It's just irrelevant on scales beyond actual fires and for questions other than fireline intensity, rate of spread, spotting, and the like. The privileging of fire behavior makes fire control the ultimate task and fuel the final referent: it reduces fire management to fire suppression and fire ecology to fuel management. That fire physics can explain some biological effects within its own conceptual dome does not mean it has the final explanations. There are other explanations, each complete within its own scope; a biological model can encompass fire physics just as readily. Fire management requires them all. When biology underwrites behavior, the sensible strategy is to turn to ecological engineering rather than shuffle hydrocarbons around a geometric space.

Everyone in the fire community will (and should) applaud a better model for crown fire initiation, plume dynamics, and long-range spotting. We need them. But they are not the basis for landscape-scale fire management. The fundamentals lie with the biology that makes fire possible at all and that shapes what burns. Moreover, few of the cultural flashpoints in fire management concern fuels; and these lie in the I-zone. The rest are typically biological, expressing a concern over biodiversity, biotic integrity, and ecological services. They will need ecological solutions. The examples of park rats, gray squirrels, and porcupine are themselves only convenient placeholders for the entirety of the living world that makes fire possible— proxies, as it were, for immensely more complex relationships, an ecological choreography, not visible to the mensurations of fire behavior.

The resulting biological models will not derive from more fundamental principles of fire physics. They will have their own foundations, and if done well they will explain why those physical principles have something to study at all.

FIRE SCARS

We're accustomed to chronicling fire history through scars. Sometimes those scars are on people.

FIRES LEAVE SCARS. They burn into cavities in many conifers. They scorch growth rings in grass trees. They brand forests into patches of even-aged stands. For the natural sciences fire scars inscribe a chronicle of fire's regimes. Eventually, trees grow over the scarred trunk as new cambium curls around the injury and forests seal off with regeneration the seared holes blown in them. Less well appreciated is that fires similarly scar people.

Burns are among the worst of injuries. Deep burns heal poorly, permanently disfigure, and may require grafts, and for these reasons the modern rules of war ban many incendiaries. But fire management cannot banish burning: its members must meet the flames, and for a few this will mean burns. Some will know burned skin, ever livid and visible; others will scorch internally. Both types of burns track the history of fire's management, but the psychic injuries are rarely visible. They are the scars borne by those civilians who lost loved ones to flames, by those firefighters and crew superintendents who lost comrades, by those fire officers who lost prescribed burns that flashed through communities, by researchers, activists, and residents who lost a landscape they were deeply attached to. For them, as for nature, life goes on, but unlike with nature, the regrowth cannot disguise the losses. The scars remain livid and deforming.

Victims brood, they pick at the defacements, they revisit the burn over and again. It's a macabre record of America's relationship to wildland fire, and one no documentarian has sought out. Those who died in flames have been honored with plaques and memorials. Those who were scarred but lived must cope by themselves. Their presence can be painful, not only for themselves but also for those who must look and pretend not to see. We honor those who have died in fires. We don't know what to do with those with disabling scars.

SMOKEY'S CUBS

Time to move on, Smokey. Take a dignified retirement. Let your cubs carry the shovel—and the torch.

WHAT SHOULD WE DO with Smokey Bear? He is now in his seventh decade and showing his age. His message has long passed its shelf life. For 50 years frustrated critics of American fire policy have sought to eject him from the community and move the American way of fire from simple suppression to a more pluralistic management. One might as well chisel away at Mount Rushmore. Smokey's old message needs to go but most of the public doesn't want Smokey himself to leave, certainly not in shame.

He was created during World War II. The Wartime Advertising Council devised the country's first national forest fire prevention program in 1942 amid alarms of incendiary attacks from Japan and fifth column arsonists. When Disney Studios released *Bambi*, Bambi became the face of the program, targeting kids rather than Axis commandos. Then Disney refused to allow Bambi to act as front man. A committee created a cartoon alternative under a stringent set of guidelines. He would have "a short nose (Panda style), brown or black fur, with an appealing expression, a knowledgeable but quizzical look," perhaps wearing a "campaign hat that typifies the outdoors." He could not resemble cartoon bears used by the Boy Scouts, Piper Cub airplanes, or political cartoonists like Cliff Berryman to represent Russia, or even the bears on a popular Forest Service bookmark. In 1944 Albert Staehle gave those prescriptions an image. Two years later Smokey got his radio voice; a year later, his tagline, "Remember . . . Only You Can Prevent Forest Fires"; and in 1948, his modern rendition from the pen of Forest Service artist, Rudy Wendelin. Smokey joined the baby boom.

They grew up together. Smokey became the Boomers' wildland teddy bear. In 1950 an orphaned bear cub was found after a New Mexico fire and became Little Smokey as life imitated art. By now the country had entered a Cold War on fire, conflating the two red menaces. The Wartime Ad Council became the Ad Council but continued as Smokey's promotional arm. In 1952 Congress passed the Smokey Bear Act to regulate the commercial use of Smokey. Little Smokey's lair in the National Zoo acquired its own zip code. Smokey had a Saturday morning cartoon show. A 1968 survey identified Smokey Bear as the "most popular symbol" in American culture. The poet Gary Snyder even wrote a Smokey Bear sutra in which Smokey stands for the defense of the natural world against the toxins, wreckage, and general insults of modern society, armed with a shovel and his war cry. "Drown their butts. Crush their butts."

Magic had happened, no one quite knew how, but they were sure they didn't want to break the spell. Yet the fire community did know the message was not true—had never been true. In fact, by removing good fires as well as bad ones the fire establishment, led by the Forest Service, by now indelibly identified with Smokey Bear, the program had caused considerable ecological damage and was creating conditions that would become economically unsustainable as well. By the early 1960s critics campaigned to put good fire back into America's wild and working landscapes. To the reformers Smokey morphed from a Santa Claus figure to a Darth Vader clone. Twenty years after Rudy Wendelin birthed him, critics despaired over Smokey. They attacked him. They mocked his message. They seemingly concluded that bad fire policy had happened because of malevolent advertising, and that removing the symbol would reform the message.

Meanwhile fire policy had been rechartered on the ambition to restore fire. When Little Smokey became decrepit in 1975 (dying two years later), instead of burying him with honors and moving on, the program found a new cub to take his place. The message continued. By then the National Park Service was nine years into the new policy and the Forest Service a year away from a wholesale reformation to accept it.

There now exists a serious disconnect between the beast and his burden, between Smokey's message and the nation's fire policy. Critics continue to hack away at Smokey as the symbol of what is wrong—his notoriety assures them attention (so they believe) for their own countermessage. The fact is, attacking Smokey has gone nowhere. It has gone nowhere for decades. A long, big generation grew up with Smokey. They dismiss the mockers and besmirchers and activists. They like Smokey. Probably they are willing to have his message rewritten. They aren't interested in trashing Smokey himself. In reality, the bear and his tagline parted long ago. Smokey stands for the benign woods and those who would protect them, and for many, a remembered childhood (interestingly he's less well known to youth today). To most people, he's a celebrity—known for being known, associated with fire protection, but not identified with any nuances in fire policy. He assures by just being there. He's less a protector from fires than of childhood memory.

Still, he needs to go. But how do we gently nudge so emblematic an icon off the stage? We formally retire him. We give him a gold shovel and a hearty handshake and the thanks of a grateful nation for doing the job we asked of him. Then we let him trundle off to a cabin in the woods and let his two cubs, the next generation, take over.

The cubs have been in Smokey's images for a long time. They appear by his side, he tells them about nature. Let them do the two jobs we need now: to fight fire and to light fire. Let a proud Smokey watch them take on the vexing fire scene of the future. Let him wax a bit nostalgic, maybe—"it was different in my day." Let him retell his story to his grandcubs, his reading glasses at the tip of his nose, reaching to temples of graying fur, and wish the youngsters well. But move him from the National Zoo to the National Museum of American History. He deserves to be remembered as a lively part of the national culture. Let him live in the past in a place where the past itself lives. But let him leave.

THE CASH-VALUE
OF FIRE HISTORY

"You must bring out of each word its practical cash-value, set it at work within the stream of your experience."
—WILLIAM JAMES, *PRAGMATISM* (1907)

So why should we care about history? What can it do, really? And when should we leave it in the cache?

B Y TRAINING AND TEMPERAMENT I'm a historian. But by formative experience and long association, I'm also a member of the fire community, and have sought to reconcile those two guilds. It isn't often easy.

My experience with the fire community is that most practitioners share Henry Ford's famous dismissal of history as "more or less bunk" and for the same reason. They look to the future. They think with their hands; they must respond to events outside their control; they must anticipate the future's cone of possibilities, not the past's wake of events concluded. Their historical horizon rarely extends more than three years. Last year, this year, next year.

When they do turn to history, they do so for practical wisdom, as a depository of utilitarian knowledge. In particular, they look to history to satisfy three needs. They want data, they want lessons, and they want meaning, and they want it all in a form they can use. They want, in brief, to convert "history" into what the great philosopher of Pragmatism, William James, called the "cash-value" of practice.

By their own training, fire practitioners believe that fire management should be a branch of applied science. So they look first for history as a source of data that they can insert into models and prescriptions. If cyberspace can be data mined, why not the past? In more cartoonish moments they might imagine historian-miners trudging off to dank archives like the Seven Dwarves, whistling while they work at prying out gems of wisdom.

The sad fact is, historical records were not written to satisfy existing models, and they can rarely provide the ready data that the fire community would like. Typically, there is too much or too little, and most of what is preserved is in a form that doesn't suit the I/O portals of software programs designed to process the output of controlled experiments. The issue is not simply that historical ore is refractory but that it's hard to distinguish the precious portion from the gangue. So while history is surely experimental, it is hardly controlled; and while it is sometimes possible to smelt that crude ore into more refined matter, the more usual response is a shrug. The stuff of history is dismissed as anecdotal. Its cash-value is suspect or worthless.

Still, if not data, history can help convey an appreciation that the landscapes they confront are not the result of abstract principles randomly stirred together like the ingredients for pancake batter. The scene looks the way it does because those pieces came together in a particular order, and not a necessary one. No one versed in the history of a place will appreciate that the scene cannot be re-placed (as it were) simply by identifying the critical pieces and processes and beating them together in any sequence; such would be a recipe for disaster. The firescapes before us are historically constructed. We ignore their idiographic histories at our peril.

―――――――――

If data doesn't work easily, then perhaps lessons might. Isn't history mostly stories? Aren't we supposed to learn from experience and draw lessons from the past? Aren't stories the best medium for preserving lessons? Such "lessons," however, are understood in a peculiar way, as part of a technological program in which experience is used to refine tools and sharpen behaviors that function as a tool. History, that is, is imagined to improve our practices and prescriptions in the same way that experience introduces continual improvements into the design of an automobile's U-joint or the protocols of open-heart surgery.

In this regard experiences are deemed interchangeable and universal in the same way that a faulty spark plug or poorly tied diamond hitch are independent of the life history or psychology of an automobile driver or a mule packer. They are testimonies not tied to temperaments. The U.S. Forest Service has even gathered volumes of such lessons from which the names of individuals have been erased. "Lessons" thus resemble "data" in that they exist apart from the actors who create and preserve them. Such a notion will seem odd to historians, but it illustrates again the extent to which the square pegs of a text-based historical scholarship don't fit into the round holes of quantitative models and the demands of legal and bureaucratic schemas.

Another trouble with lessons, as with data, is that not all of them are equal, and they are simply too abundant. It is possible to assimilate dozens of fireline experiences, but not hundreds, and as the web now makes possible, thousands. Lessons fly out of history like embers from a crown fire. There must be a process for filtering, vetting, and editing. Otherwise the past becomes a jumble, or in this case, a digital junk yard in which one might, with persistence, find a rear bumper that will fit the 1936 Ford coupe that one wants to restore, but reduces historical scholarship to antiquarian hobbyism or vocational gossip. Lessons don't by themselves, or when injected into other disciplines, make sense of the past or have the past makes sense of the present. That requires judgment.

There are arguments for checklists but they cannot be too long or intricate or specific. It's possible to remember the Ten Commandments. It's not likely anyone will memorize the unfolding litanies of ritual protocol and prescriptions that fill Deuteronomy. So, too, it's possible to recall the Ten Standard Orders, but not the metastasizing rosters of potential risks or the individual stories that proliferate like tweets. We need a checklist of stories. We need judgment to match stories with lessons, and lessons with probable fireline experiences. Piles of stories do not make an informing narrative any more than jumbles of words make an essay. Even when organized, words may add up to a dictionary, a useful reference, not a working protocol. We remember better through stories, but flashing through reams of stories can become banal and unseen, like highway billboards that finally merge into a rushing blur of the past.

═══════════

This leads to the third expectation, that history can create meaning. Instead of pretending it is a social science or shoehorning it into a technological matrix, this vision accepts—encourages—history's status as a scholarship that deals with values, beliefs, personalities, and idiographic events, and with evidence that doesn't come from controlled experiment, which is to say, it accepts history as part of the humanities. Historians preserve and celebrate the deeds of the clan. They act as chroniclers and court poets.

The past becomes usable, that is, not just as data sets or scrolls of lessons but when it becomes informed by judgment. Historians add value when they speak to those issues of ethics, aesthetics, narrative, and perceived understanding of the world that do not reside in the sciences and in fact can help place those sciences within a social and intellectual setting. They provide meaning by comparison and context. They replace certainty with contingency, and a false positivism with pragmatism. They furnish to policy and practice a historic range of variability, or what might be considered as a cone of plausibility.

The concept of a usable past is an old one, not more background babble from postmodernism. It recognizes that just as scientific models depend on boundary conditions, so do narratives. Depending on how it is framed, the same information can yield different outcomes. There is no absolute truth, only working understandings that depend on how you construct the question, how you choose to cut your slice of time, how you select your narrative voice. How you start and end determines your narrative arc, which is to say, your theme and the power of the insights and conviction it conveys. If you start the American fire story in 1492, you get one narrative, and if you begin in 1910, another. If you start in 1960, you get something again very different. There is nothing new about this idea—Aristotle laid it out more or less completely in his *Poetics*.

Those choices—the way a narrative gets created out of the raw "blooming buzzing confusion" of life, as William James put it—is the added value that historians bring to the table. By enriching our understanding of the options, by making visible the variability of history's range, they bring an informed judgment that helps sift through the past and turn it into usable data, lessons, and narratives. Meaning is not something you pluck out of the past like nuggets. It is made. It's not the provenance of professionals: it's what we all do with our experiences. The value of scholarly history is that it brings a richer sense of context and philosophy. It stands to vernacular life as the Missoula fire lab does to my backyard burn pit.

I think the American fire community understands and, within limits, welcomes this role for history. With equal measures of pride and perplexity, it recognizes that the most influential text published within the past 25 years did not come out of field or lab but from a book-lined study—a meditation written by a professor of Renaissance literature at the University of Chicago about a forest fire that happened in the Northern Rockies in 1949. Norman Maclean's *Young Men and Fire* (1992) helped connect wildland fire to the larger culture and forced the guild of practitioners to confront how they should deal with it. In Maclean's example the chroniclers and court poets found their voice, for he managed to silence the hall, and then to inspire those who heard him to do their work better. In the quest for cash-value, Norman Maclean won the lottery.

Not all historical scholarship pays off so grandly. Much is mundane, or abstract in ways that might appeal only to the guild of professionals. But the same can be said of science. Most studies add little; their real value may be that they help maintain the community of scientists. Some speak only to a handful of specialists. Humanity, after all, has successfully managed fire without laboratory data and mathematical models for all of its existence as a species. (In truth, one might identify the breakdown in fire use as the moment when a self-conscious forestry pushed aside all fire knowledge that was not vetted by formal science.)

The two genres of scholarship follow very different logics. Humanities grow only a little along a cambium fringe. Most of its literature is deadwood, or more properly heartwood, not living but still essential to the structure and physiology of the scholarship. By contrast, science is a flaming front, sometimes wide and sometimes narrow, but active only along that spreading perimeter. Its past is ash and embers. It's possible to read Aristotle's *Poetics* for insights into the nature of tragedy. No one would read his *Meteorologica* as an introduction to earthquakes.

Yet in the end science verifies data, while the humanities verify meaning, and it is meaning—that most vaporous of concepts, that least commercial of enterprises—that will ultimately guide practice because we must judge what we do by what we value, and we value only what we can endow with meaning. That's worth real money.

FROM STORY TO HISTORY

Every day's a writing day. But what prescriptions apply? How do you kindle a text that is supposedly evidence-based history out of a thicket of anecdotes, gossip, and hearsay?

EVERYWHERE I WENT in Florida I was told that written records did not exist, and that, if they did, they probably didn't matter. If I wanted to know what had happened in the past, I needed to talk to Bill or Johnny or Jody. They were the oldest participants still around; they would know. Often, "the past" was not very long ago. The old timers might have been around for 8 years, or maybe 13, or in a few extraordinary cases, perhaps 20, maybe all the way back to the medieval late 1980s. Mike and Joe would recall events according to when Harry or Pedro had arrived, or when John had retired, or sometime after that big fire had burned in the hammock or when they flew the first Premo Mark III–equipped helicopter. Maybe there were documents, there probably were, but they didn't know if there were any or where they might be located, and would have to talk to Debbie or Paula, who would call someone else. One person led to another, every attempt to explain a tidbit of fire ecology led to a story about this researcher or his quarrel with that researcher or how some grad student managed to stick a swamp buggy. The paperwork got in the way. A paper trail only led to a trash can.

It was frustrating. To most people history is synonymous with story. Clearly, I should simply listen to stories, and then write one. And stories are personal. One story, or storyteller, leads to another, as though history were a kind of oral LISTSERV. When I searched for documentation or other formal evidence that someone else could use to confirm the record, I got folklore instead. But it was quickly clear that the history that mattered was not the annual narratives stored in filing cabinets but the stories people carried in their heads. Here, fire was personal. And it could not be captured by surrogates in the form of personified institutions. That message, not caches of documents, was the lesson that filled my notebooks.

The experience challenged my undertaking twice over because it questioned both input and output. My quest for documentation—the hard record of the past—it brushed aside as spotty and trivial. My urgency about reworking that material into a functioning narrative, a historical perspective that would invest it with meanings beyond taglines, it shrugged off as academic, unusable, and probably boring. These are charges commonly laid against scholarship of all kinds, but they forced me to treat my pursuit as I might mull over the historical record. I needed to stand aside—view from outside—and ponder the unexamined life of my purpose. I had to explain how history differs from story and why this matters.

The differences are two, the first stemming from the fact that history is written and the second from the fact that it examines received accounts by rules of applied reason.

Written language is as removed from spoken as calculus is to counting rhymes. If you write like you speak, you produce gibberish. Transcripts are jumbled and tiresome; they ramble, they repeat, they misplace subjects, verbs, and topic sentences. You can write in a way that creates the illusion of pure speech, but that is an artifact of good writing. It's not how people really talk. That's the first difference—one between oral tradition and written literature.

The second distinguishes between folklore and scholarship. The truth of folklore lies in the way time has turned its material over and over like stones in a stream. What remains is durable and polished. The validation of scholarship depends on systematically reducing, testing the pieces of evidence, measuring, establishing the context. Academic history is not a science, but it is a trial of sorts, with rules of evidence and argument. Unlike a judicial ruling it must convey its judgment in narrative. And narratives differ from stories. Stories have points or punch lines; narratives have an informing principle, be it an argument, a thesis, a theme, or an organizing conceit that brings all the pieces under the reins of a single author. All the elements work to advance the arc of the plot. This, again, may look like a story—may even be written in such a way that it appears to be a spoken tale. It isn't. It's a literary construction, and the less visible the artifice, the greater the art applied.

So as I trekked around sandhill and swamp, dry prairie and hammock, looking for written evidence as a naturalist might search out threatened and endangered species, I got stories instead of documents and found storytellers rather than archives. But the raw written stuff of history is frequently scarce. What troubled was the implied need to justify what usable product I might make out of it. My hosts were too polite to ask me to explain myself. But I knew I would have to explain myself to myself.

———————————

What is the value of scholarship, or of high culture generally?

Is it just an affectation of class structure, like a toned accent or the fashionable label that identifies standing in the social order, a bevy of rituals that sets a group apart for privilege? Isn't folk culture—the learned-by-doing knowledge of how to behave properly—good enough for every-day life and the work of a fire crew? Does the general theory of relativity make any practical difference in how Floridians manage their state parks or quail plantations? Didn't Floridians burn for over 5,000 years without BehavePlus, FARSITE, or the Keetch-Byram Drought Index—and put more good fire on the ground than the high-tech moderns? Aren't stories good enough for daily life and the work of most fire crews? Wouldn't long experience, a few test fires, and a lot of work-based anecdotes do? Yes, they would. The value of formal science, and of academic history, lies elsewhere.

Scholarship relies on abstraction. That's what makes it possible to connect and create context beyond an individual person, place, or event. It's what also makes academic prose, like academic analysis, often unsatisfactory and famously tedious. Fire management typically gives the sciences a pass. It's assumed that, somehow, they will produce a gadget or a protocol that will directly improve life in the field. The actual record is a lot sparser and more tenuous than most practitioners might assume, not least because they conflate science with technology and because so many studied in natural resource programs that assumed their mission was applied science. There is no such presumption behind historical scholarship. No one is arguing for an applied humanities in fire. No one is making combi-tools in a history lab or out of English texts.

Yet the written word has its place in fire management. The stored documents help: they make possible chronologies, allow us to understand better what happened and why, and provide the raw word stock out of which can come more, better crafted words. When acted on by the discipline of good writing, a genuine narrative can move a story beyond yarns and memories and enlarge its sense of comprehension and even endow it with moral urgency. To those who doubt the value of such labors, I offer two words in reply: Norman Maclean. His meditation of the Mann Gulch fire had more impact that anything that has come out of the tens of millions of dollars invested in fire science. He showed why fire mattered.

Art, it is commonly said, comes from art, and so, too, formal history comes from other history. A student may even be made to study historiography, the history of how history has been written. Because I originally came to know fire in the field, I go to the field first and try to craft a history from what I find there, shaped by what I can later track down in libraries and archives. I spend a lot of time listening to fire folks talk. I always learn; I'm never bored. But retelling their tales is not history, which requires us to stand outside ourselves and look both around and behind. It allows us to see beyond the voice and vision of ourselves and our tribe and our guild. It means putting the stories into print not as a transcriber but as an interpreter, and ideally, as something of a poet.

E pluribus unum. From many, one—so the national motto proposes, and so many assume formal history makes a grand narrative out of a multitude of stories. But narratives are many and messy. What they promise is not a singular truth but a thematic organization and a reasoned judgment that acts on many kinds of evidence to transmute them into an order that is something more than a sum of those shards. Such an outcome won't satisfy those who demand that history should either be a science or be dumped; but in a democracy of knowledge—a republic of storytellers—it's a way to create settings for those stories to sit next to each other and hear one another's tales.

The stories I'm enjoying won't get into the narrative. But the world they inhabit should.

PROMETHEUS SHRUGGED

Megafires haven't found their Ayn Rand. Yet. But the conditions seem promising.

I T'S AN ECONOMY IN WHICH 1 percent of the population seemingly rules the rest. In fact, the proportion is closer to 0.1 percent. Not only does that tiny fraction dominate, but it is the part that has grown most vigorously over the past two decades, commandeering the most resources, distorting programs, sucking the rest dry. A small population feasts on ever more, while the rest, at best, get the table scraps. It's a galloping inequality that is profoundly deforming the national scene. And, no, this portrait does not refer to America's human economy. It refers to its fires.[1]

Those of us accustomed to understanding fire history and dynamics as the interaction of nature and culture will not be surprised to find parallels between fire history and human history. Societies have fire regimes that reflect their values, institutional choices, and collective personalities. Firescapes are as much an expression of cultures as skyscrapers, best-selling novels, or museum-housed paintings. The alignments are rougher—nature has its own logic and agency that steel girders and framed canvas lack. But one has only to compare how different nations manage fire in similar biotas to see how variously firescapes can shake out. Alaska's Yukon Valley displays a different dynamics than Siberia's Yakutia or Sweden's Norrland. The fires blackening Portugal's Trás-os-Montes look nothing like those in Italy's Tuscany, and neither resembles those roaming California's South Coast, though all are within comparably

Mediterranean biomes. They reflect decisions, both those taken by deliberation and by default, about how people will live on the land and what kinds of fires they will accept.

The fire economy of contemporary America looks the way it does for reasons similar to America's economy overall. It's doubtful that Americans (even its fire officers) consciously chose to have a fire scene dominated by what they have come to call megafires any more than most citizens chose to replace a middle-class democracy with a Wall Street plutocracy. But that is what has happened. In most of the country America's middle landscapes are shrinking and taking what might be termed the middle methods of anthropogenic burning with them. How this occurred is amenable to analysis, even if it happened through the laws of unintended consequences. What it means for the future may well occupy the coming politics of wildland fire.

—————————

The consensus perspective on how the American economy morphed over the past 40 years points to globalization, an internal shift from manufacturing to services, and policies that aimed to liberate free-ranging capital from restraints and regulation. Eerily, analogous concepts can apply to its fire scene. For globalization, look to the global change suite that includes climatic warming and exotic invasives. For the leap to services, see the movement of private land from commodity production to amenities communities; of public land to wilderness and nature protection; and public sector investment to private sector capital. For policy, note the determination to restore free-burning fire to something like its earlier untrammeled state, while removing firefighters from heedless harm and protecting the McMansions and recreation-home assets of America's privileged classes. Appropriate management strategy promises to reinstate a more laissez-faire order, what might be termed a trickle-down fire ecology, on the assumption that more fire will inevitably improve the functioning of ecosystems on a landscape scale. It's a faith-based environmentalism. All it lacks is an Ayn Rand to craft a manifesto in the form of a novel.

There is little doubt that America's fire scene around 1970 was too hegemonic, too shackled by rules and an assumption that control must precede any loosening of norms, that it had become both too single-minded and too cautious, even sclerotic, in practice. Too much of American fire lay under state sponsorship. The fire economy needed opening up. It hungered for a more pluralistic, experimental structure of governance.

But it's doubtful that the contemporary scene is what fire officers and strategists expected to replace it. That the wildland-urban interface was identified as a geographic entity and minted into its own neologism in the mid-1980s deftly aligns the timing of the two economies, nature's and the nation's. It also captures exactly the growing bifurcation into two antagonistic groups that have progressively defined the American firescape. Between them they have abraded away not only the middle, working landscape but the middle methods by which people traditionally lived on them. More and more on the public lands fire was either suppressed or left to free-range as much as possible.

It's doubtful, too, that one might expect a protest movement to Occupy the San Bernardino or the Payette. The National Cohesive Strategy for Wildland Fire Management is a shrewd assessment of current trends, but its primary lapse—not part of its charge—is its inability to show how fire management will valence with the larger culture. This matters because the two economies, fire and national, like their two politics, are related by more than metaphor. The polarization of one is an echo of the other. The resources and decision processes of the national scene will determine those available to manage fire locally. That observation extends also to the amount of gray matter the community can allocate to the task.

Is the new order wrong, or unwise because it derives from a belief that overregulating fire is a road to environmental serfdom? If "fire is fire," does it matter how that fire is arranged on the land, any more than how capital gets distributed among society? Does an intense concentration of burned area and costs into a fire plutocracy affect the outcome nationally? I think it does. But the country is undergoing an immense experiment, perhaps a kind of ecological shock treatment, which will test that idea, along with those of everyone else, about how we propose to distribute our assets and live with fire.

THE AMERICAN FIRE COMMUNITY'S EURO MOMENT

Analogies are like fireweed, or ragweed, depending on your perspective. Some are more useful than others. Here's one as the National Cohesive Strategy goes into the field.

THE SENTIMENT IS INCREASINGLY voiced that, amid global change, particularly a slow upheaval in the global climate, we face a no-analogue future. It's a beguiling notion, and one especially appealing to those alarmed by recent fire seasons and animated by a sense of urgency. We're sailing over the edge of history's map to a new world beyond the ken of the old. Forget triangulations with the past. History is, in Henry Ford's phrase, "more or less bunk."

But we don't need history as a source of analogues. They swarm around us like midges. The swamp of the future will be as full of them as mosquitoes on a summer day. The question is knowing which of them speak in helpful ways. In contemplating the tortuous state of the American fire community, my personal favorite is the slow smash-up that has been unfolding in the euro zone.

The euro proposed a monetary community without a commensurate political union. Countries from Malta to Finland, Greece to Netherlands would share a common currency overseen by a central bank with limited powers. For countries that adopted it the euro was a powerful fiscal stimulant. It instantly extended access to high-value funds even to countries that lacked an infrastructure adequate to absorb them, that were more ready to consume than invest. When the money tightened, the stresses became unbearable and the system cracked.

The outcome has been to divide the euro region into two realms, one central and one peripheral. As the flames of fiscal crisis have spread, various oversight agencies have sought to douse them with emergency funds. But the feral fires keep returning, the bailouts get larger, and donors have become less forgiving and hopeful. They want serious structural reform to work down the accrued debts and end the demand for further tranches. Some critics demand austerity as the only answer. Others argue that growth is the best solution, and this requires fundamental changes in the recipients' economies, which is to say, in how they live. Meanwhile, the contagion has continued to spread. It is one thing to have Greece crash. It's another to flush Spain.

Among the cognoscenti the sense grows that the euro realm must either reconstitute its governance structure to align with its fiscal, or else it must break up. The alternative—a Sisyphean exercise in crashes and bailouts—cannot continue. Those with the money don't wish to pour it down bottomless sinkholes, and those who receive it are unable to leverage it to escape their accumulated debt bondage. The money is simply spent. The problem persists.

———————

Anyone familiar with the American fire scene will recognize the signs. Over the course of a century, a system of interlocking agreements was negotiated that cobbled together fire agencies into something resembling a national union. The U.S. Forest Service organized the matrix. Eventually every state, large and small, every federal land agency, the gargantuan and the gangly, could assist every other. In practice, the big supported the little. By 1970, however, even giants like the Forest Service could no longer operate on their own and had to turn to others for support during emergencies. "Interagency" became the vogue term, and "total mobility," the capacity to freely move crews and engines to where they were needed, became the Schengen agreement for American fire.

What held it all together was not good will or a neighborly desire to help, but money. In particular, emergency funds—off budget, liberated during fires—saturated the system like groundwater. There was a logic to the arrangement because no one could predict in advance what kind of fire season they would face. Ultimately, those monies came from the federal treasury. Through transfers they could move to wherever the crisis struck. They made it possible for those with smaller fire economies to join those with bigger. They glued an odd-bodkins bin of fire institutions into a pastiche that resembled a system.

In practice, the system worked, even if in principle no one liked it. It would be far better to have budgeted funds from a controlled spigot that could be spent on prevention and presuppression than to simply open hydrants during emergencies. In 1978 Congress severed Forest Service authority to access those supplemental funds while at the same time boosting its base fire budget. Nearly a decade of relatively quiet fire seasons followed. Then the West entered a long drought. The Yellowstone fires of 1988 blew any restraint away. Year after year, somewhere, fires roared, and increasingly they pressed against the exurban communities that were reclaiming rural America. Faced with telegenic emergencies luridly broadcast on the evening news, Congress ignored its own rules and paid for the emergencies. Rather than being weaned off supplemental funds, the monies became addictive.

The 2000 National Fire Plan boosted budgets for preparedness, but instead of watching the flames quell, the West coughed up a decade of megafires. In 2003 the Bush administration decided that firefights in Montana and California were less significant than those in Iraq and Afghanistan; the Forest Service was told to stay within its programmed budget. Fighting the monster fires consumed nearly half of the agency's funding, and it was no longer able to cover the costs of its cooperators. The system plunged into massive deficits, only redeemed at the price of a crippling austerity for everything else. So even as the fire scene quickened, the system entered a slowing spiral in its capacity to respond. Like Wall Street's house of securitized cards, the American fire community's overleveraged mutual aid agreements threatened to collapse and take the entire system with it.

In 2009 Congress passed the Federal Land Assistance, Management, and Enhancement (FLAME) Act. FLAME was a kind of Troubled

Asset Relief Program for fire: it was intended to arrest the free fall of big-fire deficits. It boosted the base budget for firefighting, enough (it was believed) to allow some fiscal control over operations. The expectation was spelled out that agencies should better anticipate their needs and "prevent future borrowing from non-fire programs." FLAME would end the fiscal transfers. But the fires kept coming, Congress failed to fund its own program, and the contagion spread from California to Colorado and Montana to Texas. The fires are no longer banished to the periphery. They are moving into the core.

They have burned away not only the new monies but any pretense that further supplements will not be needed. The Office of Management and Budget cannot dictate fire seasons. The fact is, the American fire community is woefully undercapitalized. The critical landscapes are too big to fail and too large to manage. In May 2012 the Forest Service reversed 44 years of policy reform and in the name of a fiscal emergency reinstated the old all-suppression policy. Better to attack every fire quickly than risk long-lingering burns that could run up costs. Instead of fixing the problem, the new funds, like another bailout package to Greece, have only calmed until the next flare-up.

———————

Embedded in FLAME was a requirement, based on a notion first broached by the Government Accountability Office a decade earlier, that the agencies devise and submit to Congress a National Cohesive Strategy for managing fire. The idea was to better coordinate programs and so fill lapses and shrink overlaps. But that proved a matter of politics, not just policy.

The American fire community had created an arrangement for sharing resources, backed by emergency monies and other subsidies, but without a political structure by which to govern among all the unequal parts, which now had to include states, counties, cities, volunteer fire departments, fire districts, nongovernmental organizations, and private landowners, all of whom had different purposes, capacities, and risks. The American fire community had to reconcile very different economies of fire and distinctive historical geographies, some deeply rooted. The upshot was, the AFC could spend money but not oversee the conditions that mandated those expenditures. Ultimately, you control fire by controlling the landscape in which it burns. So, too, if you wish to control fire's costs and reduce its damages, you must control the institutional landscape within which those fires happen.

Discussions that underlay the National Cohesive Strategy began in October 2008 with an informal congress of fire officers at Emmitsburg, Maryland. The Emmitsburg 13 realized that it was impossible to manage the national fire economy without rechartering the politics that underwrote it. They would have to bring together the frequently cooperative but oft-suspicious and always factious members of the American fire community to reconsider roles, rights, and responsibilities and to draft a new constitution for American fire—or rather, to move from an Articles of Confederation to a federal constitution. A loose union for transferring firefighting resources and funds had to be replaced with a more orderly mechanism for allocating monies and attention. The Emmitsburg 13 issued a series of foundational documents to direct the process. They openly described their proceedings as a kind of *Federalist Papers*. Then the FLAME Act mandated a formal program. The NCS completed phase 1 in March 2012. We are now well into operations. The NCS has no funds and no legislated power. It relies on the community to reach consensus.

The National Cohesive Strategy is happening because the FLAME Act ordered it. But behind it lie two competing theories of how to reconcile fire and finance. One strategy promotes austerity. The fiscal transfers must end; the agencies must live within their much-reduced means; and if the land burns, they can rebuild landscapes out of the ashes. The other strategy wants significant stimulus to get ahead of the fires, to build resilience into wildlands and exurbs so the system won't incinerate and won't require endless transfers in the future. But neither can impose an ecological order unless the political order is in place to make choices that the community recognizes as legitimate about what to emphasize and how much to spend. It's about governance, not just policy. Ultimately, it's about reforming how we live on the land.

With the NCS so far advanced, the American fire community is ahead of the euro. A deeper integration is underway at least on paper, although it is unclear whether the fire constitution will, when final negotiations are complete, be accepted. The process has both carrots and sticks, large and small. Some places and programs will gain funding, some will lose. If the American fire community fails, the threat is implied that responsibility for fire will be turned over to the Federal Emergency Management Agency, which would be the equivalent of having the International Monetary Fund run the euro zone. While that threat is a political rant, not a reasoned policy, it helps place fire on a Richter scale of attention. Strengthening the analogy, too, is the realization that there is no Plan B. If integration stalls, a fragmentation of the American fire community is the logical outcome. The rich landscapes will take care of themselves. The poor will burn.

Fix the deep politics or break up—that is the challenge before the euro zone. And it is why now is the American fire community's euro moment.

THE WRATH OF KUHN

A fire triangle for understanding the ways we understand fire, and why we need all three facets.

W HAT SHOULD FIRE research study, and how?

The traditional answer is that fire is a reaction in physical chemistry. But when the deep driver of planetary fire is the burning of fossil biomass, when free-burning fire is mostly disappearing from the developed world except on reserved lands, when public concerns over nature protection hinge on biological indices like ecological integrity, biodiversity, and sustainability, when the prime threats from wildfire are to houses along a fractal exurban fringe, when the common element among every fire problem is humanity—people as fire kindlers and fire suppressors, people as direct and indirect shapers of landscapes on a geologic scale, people as judges of what is and is not a fire problem (or a solution), people as operators of industrial combustion on such a scale that they are unhinging even climate, that Ultima Thule of fire's physical environment, then the contemporary design and emphasis of fire research might well appear to an unbiased outsider as the residue of a cockeyed historic evolution. Like the drunk who keeps searching for his lost keys under the streetlight because "that's where the light is," fire research continues to elaborate a physical paradigm because that's where the funded science is. But the keys to understanding may lie elsewhere. The time has come to recharter our conception of what fire is, how we might study it, and how we ought to manage it.

Consider a research program that arises from three conceptual constructions of fire. Each is internally consistent, each fully encompassing, each equally necessary, each in its intellectual power coequal with the others. They are the equivalent for fire research of non-Euclidean geometries. The historical reality, however, is that one ring has ruled them all.

THE PHYSICAL PARADIGM

This, the founding paradigm, asserts that fire is a chemical reaction shaped by the physical characteristics of its environment. Those physical parameters shape the zone of combustion as it moves about the landscape; how that happens and with what consequences defines the realm of fire research. These are fire's fundamentals.

From such processes flow all other fire effects, and from this model derive all other explanations for why and how fire exists on Earth. Fire ecology is the study of how fire, as a physical disturbance, interacts with the living world. Fire policy and fire sociology are the study of how, granted fire's physical properties, people should apply and withdraw fire and how they should protect themselves from its threats. Fire management consists of exploiting fire's physical behavior—to check its spread with physical countermeasures and to kindle its benefits by arranging ignition and fuel. Fire science is the study of landscape as firesheds, its combustion chambers framed by weather and terrain. The future of fire research is to extrapolate the physical paradigm into more and more phenomena.

This conception does not arise inevitably and uniquely out of the subject. It expresses the bias of past funding, which reflects the desires of state-sponsored forestry to control free-burning fire on public lands. These agencies typically assumed exclusive sponsorship over fire scholarship. Without them it is possible there would have been almost no research done, but through them funding privileged some topics and disciplines over others. The issue, in brief, is one of intellectual politics and institutional monopoly. Fire never got properly situated in biology, the social sciences, or the humanities. Research flowed over a single floodplain, and the more it flowed, the more deeply the dominant channel entrenched itself. Even recent initiatives like the Joint Fire Science Program in the United States (welcome as it is), demand a scholarship based not only on natural science but on those branches of science capable of translating ecosystems into fuels and describing how fuel arrays influence fire's behavior.

The problems with this approach are several. Partly they arise from the limitations of this perspective relative to the problem at hand, partly from way the physical paradigm is imagined, and partly from the tendency to demand that all other conceptions align with this paradigm in what becomes in fact (if not by intention) a hegemony. Laplace famously (if fatuously) pronounced that if he knew the position and velocity of every atom he could predict the future of the universe. As elaborated, the physical paradigm echoes this claim, that once it knows the position and flammability of every fuel particle, it can predict the outcome of every ignition, that all it needs to complete its public agenda are better models, denser instrumentation, and more powerful supercomputers. This is less a hardnosed philosophy of science than an archaic one, long abandoned in higher physics. Even if it could be done, controlling fire's behavior is only a fraction of fire's management.

Still, this would not matter much—each discipline could search out its natural level—if the scene did not come with the implicit understanding that what doesn't fit these strictures doesn't exist, not as science. Other scholarship acquires power only to the extent that it flows from or aligns with these precepts and style. The issue, that is, pivots on the privileging of the physical paradigm over other conceptualizations. During the 1960s the Tall Timbers Research Station offered an alternative forum, outside government funding agencies, for promulgating ideas about fire ecology and prescribed fire. In today's political lexicon it contributed to a more textured civil society. What the fire community needs now is an intellectual equivalent, for while it has become a truism that the sticking points in fire management involve ecological themes as refracted through cultural understanding and politics, the discourse must be cast in the language of fuels and limited to questions of fire's control, which is to say, it must remain within the domain accessible to the physical paradigm.

Presently, the physical paradigm so suffuses fire research that it seems less an intellectual convention than an axiom. It is everywhere, and its accomplishments are genuine—undeniable, ubiquitous, and profound. It continues to produce intellectual excitement as well as practical prescriptions, of which recent studies on crown fires and eruptive fires especially stand out. Most of the fire community can probably imagine no other approach. Yet like Euclid's Fifth Postulate, the physical paradigm is neither inevitable nor logically necessary. Change its founding assumption, and other, wholly consistent conceptual geometries of fire are possible.

A BIOLOGICAL PARADIGM

An alternative perspective might consider fire as biologically constructed, as a reaction created and sustained by biological processes. In this conception, the fundamental conditions of fire's behavior are set by the living world: life is why fire exists, the living world molds fire's expression, and physical parameters matter only insofar as they are refracted through a biota. Fire's environment is primarily organic. Fire's real fundamentals reside in the properties of its biotic setting.

The basis for such a proposition is simple: life creates oxygen, life creates combustibles, and life, through the agency of humanity, overwhelmingly creates the sparks of ignition. The chemistry of combustion is a *bio*chemistry: fire takes apart what photosynthesis puts together. Within a cell, the process is called respiration; within the wide world, fire. The arrangement of combustibles is fashioned by evolution and ecology, which are themselves responsive to biological processes, not simply derivatives of a physical environment. While terrain and climate help shape particular patterns of fire, as they do for evolution and ecosystems generally, they express themselves through the resulting arrangement of organic matter, or biocombustibles. Fire's integration occurs within the biosphere.

In this conception, fire's ecology is not simply the record of disturbance by mechanical forces acting on a biological medium but a propagation *through* a biotic medium. Wind, ice, debris flows, floods—all can occur without a particle of life present; fire cannot. It literally feeds upon biomass and more resembles an outbreak of bark beetles or SARS than a windstorm or a glacier. The expression that such-and-such a disease spread like wildfire could be restated to read that such-and-such a fire spread like a disease, a contagion of combustion. Life need not simply adapt to fire: it breeds, nurtures, and shapes fire. Fire becomes less a mechanical force that impinges on ecosystems so much as an organically informed process that manifests itself in such physical expressions as heat and light.

Ideas have consequences. Overall, fire ecology defers to the physical paradigm, which strikes the data of ecology much as it imagines fire striking an ecosystem. At issue is not whether fire is "natural," but in what way it is natural. Is it a physical process like wind and lightning that must, in its basics, be described by physical science? Or is it an organic process that cannot be abstracted from its biological context? In some ways it is both, and neither. But if the first, then ecological solutions to fire problems must ultimately lie with physical countermeasures, such as slashing, burning, quenching, and rearranging blocks of hydrocarbons. A biologically based paradigm by contrast would propose biological controls.

These could range from genetically modified fuels to ecological engineering. Instead of fuel-laden firesheds, we could imagine fire habitats that fire shares with hosts of organisms, some of whom compete with it and all of whom shape its setting. Fire's reintroduction would resemble the reinstatement of a lost species. Fire's control would reside in its biotic context rather than along its flaming front. Here is a paradigm better suited to management on a landscape scale. Here is a conceptual language, as "fuel" is not, to describe the significance of fire to biodiversity, ecological complexity and integrity, and sustainability. The physical paradigm was not propagated to answer such questions. Twisting it to do so is like asking a creature with flippers to handle objects as one with fingers could.

The breakdown, however, lies also with biologists, who have not seized upon fire for its value as a way to understand the peculiar properties of life on Earth. In recent years ecologists have broadened their niche—have, for example, begun to compare fire and herbivory; have argued for the evolutionary antiquity of fire as a selective agent; have explored the genetic heritage of flammability; have strengthened the thesis that fire is somehow natural and necessary. But these do not add up to a biological paradigm of fire that would characterize fire as fundamentally a phenomenon of life—not simply something that life adapts to but that life makes possible, whose primary parameters are organic and whose explication and control should reside in its ecological setting.

Perhaps the biggest payoff is that a biological paradigm could incorporate humanity into the grand narrative of life on Earth. If a megafaunal mammal had emerged during the late Pleistocene and claimed a species monopoly to start and stop fires at will, and had blitzed across the planet, you can bet there would be considerable interest among biologists in what this creature means to planetary ecology. There would be subdisciplines, journals, and symposia devoted to the subject. Yet just this scenario happened. Apparently because the creature is *Homo sapiens* the topic stands outside biology, or if it is considered fit for analysis, the model is, once again, that of fire crashing into biology from the outside, much as people are presumed to force their fires, unnaturally, onto landscapes. It is possible to pick up the other end of that stick, however, and argue that in reality humans are completing the cycle of fire for the circle of life. Through people, life is increasing its control over combustion. The master narrative of fire is that it is becoming more rather than less biologically informed. The incorporation of people as biological agents should be one of the prime assets of a biological paradigm.

What may surprise casual critics is that this proposed reconceptualization is potentially as coherent and consistent as the physical paradigm, and that it is fully capable of absorbing its rival. A biological paradigm can offer a parallel worldview in which fire becomes another expression of a biological Earth, along with the rise, extinction, and arrangement of species, and the cycling of carbon and nitrogen, and for which physical models become subroutines within the grand programming set by the living world. Under the prevailing regime, biological traits must be recast to fit physical parameters, not physical traits to a biological conception. Yet just the reverse is entirely possible. A biological paradigm would center fire within the living world, look to life to contain and exploit it, and make fire into something more than a sidebar and errant footnote in standard texts in ecology and life science.

A CULTURAL PARADIGM

The cultural paradigm is both the most obvious and the least developed. It focuses on fire's species monopolist, humanity, for whom fire's manipulation has always been a defining trait and whose present dominance relies on continued control over combustion. The story of fire on Earth is increasingly the story of what people do or don't do, directly or indirectly, with regard to fire and its setting. The cultural model would seek to record and explain this interaction. It assumes fire's fundamentals reside with us.

The difficulty with this approach is not that nothing has been studied, but that no conceptual organization is in place to make sense of the data. A few studies exist, mostly from anthropologists, some examining aboriginal societies, most detailing slash-and-burn regimes, a couple focusing on the fire-based agriculture in Europe and America and changes in settlement rhythms. Only a handful attempt to tinker with a general model that links the diverse aspects of human fire practices or even to sketch a historical chronicle that embraces the full narrative from lightning bolts to SUVs (academic history has almost wholly ignored the topic, save for urban settings). There is no scholarship to analyze what institutions best serve fire management. There is no intellectual history of fire after the Enlightenment. There is almost no inquiry into fire as an organizing device for the human occupation of the planet. And there is no truly *political* history of fire, which seems bizarre granted that contemporary fire agencies are overtly political institutions; for over a century the apparatus of wildland fire protection and research has occurred almost wholly under state sponsorship.

The reasons are the usual ones. There is no discipline of fire, so fire occupies nooks and crannies in other fields, and while there is funding for natural science, there is little for the social sciences and none for the humanities that deal with the institutions, social values, and cultural choices that guide humanity's fire behaviors. Such omissions become grotesque as one considers that the deep driver of fire on Earth today is the industrial revolution, which for fire history means replacing the burning of surface biomass with the burning of fossil biomass. For the developed world, the period of transformation—call it the pyric transition—is a time of abusive and damaging conflagrations. The demographics of industrializing fire closely resemble those of industrializing people, beginning with a veritable population explosion and ending with reproduction below replacement levels. Such eras became a powerful background to arguments in 19th-century Europe and North America for state-sponsored conservation and a flawed baseline from which to estimate background burning. Presently, however, there is no sense that this transformation is a part of fire research except as global warming might quicken the opportunities for megafires. A cultural paradigm, however, cannot avoid such matters: it is humanity's fire practices that are propelling global warming and humanity's institutions that must cope with the maldistribution of burning that industrialization has sparked.

Fire problems are socially constructed problems. They are problems because people define them as such, and nearly all the crises that fire inflames can be resolved by social means. After all, humanity has lived with, and exploited, fire for all our existence, and we have fashioned the fire regimes of the planet without the benefit of academic science. Trial and error, socially coded into prescriptions, have served to make fire available and keep it within acceptable bounds. People have been able to scorch fields, drive game, cook, fumigate, foster metallurgy and ceramics, counter wildfire, burn pastures, and otherwise put fire to the service of sword, plow, hoof, and hammer without a scrap of modern science. (On the contrary, Enlightenment science in the service of the modern state has systematically extirpated that entire encyclopedia of folk knowledge.) As the physical paradigm suggests physical means to control fire, and the biological paradigm biological controls, so a cultural paradigm would propose cultural controls. The available means are many, for whatever shapes people's understanding and behavior can affect how they decide to apply and withhold fire. Whatever their ultimate origins, the levers of control will operate through institutions and ideas. They will reflect how we think of fire and define it as a concern.

This observation does not mean that people can exert total control over fire, only that we have the capacity to determine what control means and what kind and how much we will accept. All the data that the physical and biological paradigms can spew out will matter little if there is no means to apply it in ways that satisfy the human definition of the problem. The integration of these factors occurs within a social setting. Equally, we can resolve almost all the urgent problems of fire management through social means if we choose—by land management, by regulating behavior, and even by deciding whether a fire free-ranging through wildlands is a cultural crisis or an ecological spectacle. Mantras that "fire is natural" ignore the fact that in the industrial world free-burning fire thrives largely on reserved public lands or on temporarily abandoned lands; neither condition is a state of nature; they result from social decisions. Even that ultimate natural referent, climate, is becoming unhinged by anthropogenic fire practices.

The cultural paradigm, that is, can absorb the other two. It can even explain why the physical paradigm dominates contemporary imagination, for science is a human enterprise conducted through institutions, and the question of which science gets supported (if any at all) is a social choice. Moreover, the cultural conception can offer its own justification for research, not for the techniques that science furnishes but as a cultural undertaking like art or opera that lifts fire's management beyond the realm of craft and guild. Understanding these dynamics matters as much as modeling the footprint of retardant drops or torching firs. American and Australian fire agencies, for example, have not been hammered in recent decades because they could not calculate fuel loads but because they could not speak to their sustaining societies. In this figuration, the physical and biological paradigms become sources of particular information that serve the cultural paradigm, which alone addresses the core issue, how should humanity use its firepower and conduct itself as a fire creature?

Over and again, the sticking point on fire policies is uncertainty over what people want a landscape to be and what means they may have to pursue those ends. Before such matters, the physical and biological paradigms stand dumb. And while a cultural paradigm could not speak with the putative authority of laboratory science, it can at least speak with the voice of scholarship, and while not declaring what choice we should make, it can help us understand better the nature of the choices before us. Technology can enable, but not advise; science can advise, but not decide. For that we need other scholarship and a conception of fire that makes the process of deciding a point of integration, not an afterthought. We need a paradigm that unabashedly places people as fire creatures at its core.

REGIME CHANGE: EXPLAINING MEGAFIRES

Over the past 20 years, the world has witnessed an epidemic of megafires. Countries that thought they had banished open fire into their ancient past have seen it return; countries that assumed they had mastered wildland fire, holding it to a quantum minimum, have seen it break out of suppression with shocking fury; countries eager to develop new lands, either as plantations or parks, whether by clearing old forest or reclassifying former forests, have seen fire and smoke spread like a plague. And all have nervously pondered the growing evidence for an era of global warming, amid the scientific conviction that industrial combustion is an unwanted accelerant. How might the various paradigms interpret this outbreak?

The physical paradigm would point to climate change and its effect on fuels. Places that were normally wet became droughty, and hence available for burning on a grand scale, while landclearing and land abandonment plumped vast terrains with combustibles. Fire simply followed those fuels and the tidal rhythms of climate. Analyze those arrays and you can understand and predict the nature of the impending conflagrations. Only in limited circumstances can people modify the master parameters through manipulating "fuels"; mostly, they must meet force with force. Megafire thus appears like a tsunami, and calls for such remediation measures as physical barriers, early warning networks, and policies to compel people to adjust their lives to an unalterable reality. Too often, however, such analysis leads to the landscape equivalent of stuffing buildings with asbestos in the name of fire protection.

The biological paradigm would point to disrupted biotas—to broken forests, invasive pyrophytes, the collapse of internal checks and balances within ecosystems. The fires have behaved rather like an emergent disease, a pyric version of avian flu, with climate helping create favorable conditions (although catalyzed and boosted by human practices), but with the propagating medium and vectors residing in the living world. What had been a seasonal nuisance has now mutated into a virulent and lethal plague. Rather than firefighting with pumps, aircraft, and retardant, a strategy of containment might look to epidemiological analogues and public-health strategies from vaccinations and sanitation to quarantines and select emergency care. The effects of such fires, even the worst fire plague, would vary enormously among the populations affected. The metrics for determining the seriousness of an outbreak would reside in biological indices.

The cultural paradigm would note that, while drought and lightning have accounted for many of the burns, it is primarily people who are the agents of the outbreaks and who have determined what responses, if any, the fires warrant. They would note that the eruptions have resulted from interactions of natural conditions with changes in land use, institutions, policies, and perceptions, all of which have created opportunities for fire, and which suggest that megafires are analogous to a revolution or an evolving insurgency. Big fires have resulted from breakdowns in the apparatus for fire control that followed political upheaval in Russia and especially Mongolia. Big fires have swept half or more the area of flagship national parks in America, South Africa, and Australia as a result of changes in policy and practice. Horrific fires have plagued Portugal and Provence from rural land abandonment, and Brazil and Borneo from subsidized transmigration schemes. The lightning-kindled conflagrations that have blistered North America have occurred on public lands; had those places been converted to shopping malls, landfills, or trophy-home suburbs a very different regimen of fire would be likely. And they occurred after policy reforms that sought to increase fire's presence, a goal they achieved though not in the way intended (be careful what you wish for).

All these burns of course share a common chemistry of combustion, all feed upon biocombustibles, and all have drought as a precondition, but none of these factors is either by itself or collectively sufficient to explain the outbreaks. The unifying catalyst is humanity. It would seem that even climate is now bending before humanity's combustion habits. The burned area racked up by megafires, which seems like a proxy of climate change, is in reality an index of climate and people interacting. That this increased burning could, simultaneously, be presented as evidence of both fire management's failure and its success further testifies to the power of fire as a cultural creation.

A fuel array, a habitat, a landscape—nature does not determine by which prism a place might be viewed; culture does. A serious fire scholarship would begin with that irreducible fact.

INCURRING THE WRATH OF KUHN

Calls for paradigm shifts are a cliché of our time, as common as the falling of autumn leaves, and anyone who invokes them deserves the righteous wrath of Thomas Kuhn. Most fire scientists will elect to do what they have done in the past. They will surely argue that the three conceptions of fire are really variants of one universal science, bubbling up from a common wellspring of data, that there is no need to parse that collectivity into a suite of thematic models, much less to burden the field with proliferating paradigms. They will insist that the nucleus of research must be fire's physical description, or else science will lose its rigor and its claim to primacy among the several voices of fire scholarship. Persistence forecasting will argue that the near future of fire research will resemble its near past.

This protest misses the point. Each paradigm can claim to survey comprehensively and coherently the whole field; each is fully capable of absorbing the others' data and formulas into its own grander schema; and there may be no way intellectually to assert a transcendent vision over them all. Among them, arguments can revolve like a game of rock-scissors-paper. Each can declare itself supreme; each can claim its core criteria as fire's true fundamentals; yet none of them is sufficient in itself. A wise fire research program would embrace them all, crafting policies and practices from each according to its abilities, supporting each according to its needs. The physical paradigm works best for firefighting and asset protection; the biological paradigm, for the macromanagement of landscapes and biotas; the cultural paradigm, for defining problems, confirming ideas, and establishing institutions to administer humanity's species monopoly over fire. The value in segregating them is that each can only realize its conceptual potential if it can follow its own internal logic to conclusion. Leaving them to free-associate or merge biases the outcome in favor of the historic default settings—the physical paradigm for research, suppression for policy, and the firefight for narrative.

Fire scholarship deserves better. We are uniquely fire creatures on a uniquely fire planet; that we invest so little in fire's study speaks to a deep disconnect between what we are doing on Earth and how we understand our behavior. Few topics can speak more profoundly about what it means to be human. Fire offers a synthesis that few other topics can even imagine: every aspect of its expression deserves far more study than we have invested in it.

Yet almost monthly the chasm widens between what the physical paradigm can do and what the fire community needs. We shape policy and practice to fit what our science can say, not adapt the science to ecological and social requirements. Thus in the United States the determination to have a "science-based" policy means that projects to restore fire must be framed in terms of "fuels management" when the issue may be ecological integrity, while arguments over social values regarding nature preserves or exurban sprawl must be translated into fuel loadings that are literally meaningless within the context of that conversation. (Consider as well the arid scholasticism that afflicts the Australian discourse over "hazard reduction burning.") The only remediations allowed are to do nothing or to reduce fuels, which is the intellectual equivalent to suppression by mechanical means. Yet the future of prescribed fire will not reside in its role as a flaming wood chipper, but in its capacity as an ecological catalyst. Similarly, the determination of suitable fire policies amid an era of planetary warming powered by anthropogenic combustion habits will not reside in either the physical or biological paradigms, and insisting that it must only wedges research further apart from reality. Too often existing programs are not really investigating the problems: they simply do the science traditional to the field, and when the results prove inadequate, they plead for more funding for more science of the same sort, and then demand that society conform to their research, not research to society. Scholarship thus fails because it is the wrong scholarship, which leaves the field of discussion to raw politics, untethered rhetoric, or the ecological equivalent of faith healing.

It is hard to see how this bias can change, however, until fully biological and cultural conceptions of fire are allowed to flourish as alternative paradigms. It is not enough to go beyond the realm of today's assumptions but to go beyond the belief that what we need to know can ever be derived from them. The elaboration of alternative paradigms should be a priority task. We need all three, for as with that older fire triangle, remove any side, and the fire will go out.

UNTAMED ART

At the Roscommon Equipment Center, I saw a Japanese version done in silk. At the Myles Standish State Forest I saw a framed print. My wife found our first copy in a thrift store in Altoona, Pennsylvania. Welcome to the fantastic saga of the world's most oft-reproduced fire painting. It's a story of wars, frauds, double-dealings, no dealings, art lost and sometimes found. But you will see the image everywhere.

THE WORLD'S MOST FAMOUS painting of a forest fire is also its most misidentified. Anyone even casually familiar with fire art, or who has shown the least curiosity about prephotographic images of wildland fires, will recognize the scene instantly. The focal point is a slow-swirl pillar of flame rising through a patch of boreal forest. The fire gathers in a ragged eddy along the forest floor before sweeping upward against the wind and twisting through the canopy with a convective heave and a reverse eddy of flame and smoke. The symmetries are nearly perfect: sky and earth balanced with a layer of mossy forest between them; the deep woods wedging to the center and there cleaved evenly by that archetypal spiral of flame.

The painting has been widely reproduced, and variations on its scene abound in assorted media, with varying internal proportions and sizes. Some versions insert fire-scarred pines; some even include a cottage and firefighters. You can find them in color lithographs hanging in kitchens, old garages, even a few bars, and bins at secondhand stores. The U.S.

FIGURE 2. This, in black and white, is the most widely disseminated version of Denisov-Uralsky's celebrated painting, cropped vertically so that the smoke plume does not show. This is the copy from which all other copies derive. The original was in vivid color, the core of which is shown on this book's front cover.

Forest Service has a black-and-white print in its historic photo collection. A Bavarian ceramics company reproduced it on porcelain plates. Pre-WWII Japan manufactured facsimiles using silken thread. Grandma Moses copied the scene, as have other American primitives. A Wisconsin woman won a folk-art festival by submitting a variant she painted, fraudulently claiming she reproduced the image not from a reproduction on her living room wall but from real-world fires remembered from her youth. Others have claimed that the scene commemorates the 1871 Peshtigo fire. More recently, versions have appeared on eBay amid various testimonies to authenticity ("original oil") and prices ranging up to $850.

All come from a common source, a painting by the Russian artist Aleksei Kuz'mich Denisov-Uralsky originally titled *Lesnoi pozhar (Forest Fire)*.

Born in Yekaterinburg into a family descended from Old Believers, his father, Kuzma Osipovich Denisov, was a miner turned dreamer turned artist, who worked with mosaic reliefs, landscapes, and icons, and made a reputation by constructing gigantic grottos from gemstones for display at Russian and international fairs. Aleksei Kuz'mich, born in 1860, was trained to this same craft and had instilled in him that semiprecious stones, not paintings, were the essence of a good career. In 1882 father and son participated in the All-Russian Arts Industrial Fair in Moscow, where the son's mosaic paintings attracted special notice. Shortly afterwards Kuzma Osipovich died, leaving to Aleksei the task of supporting his mother and three sisters. He persisted in his craft, his artisanal skills winning further attention. In 1884 he received formal designation as a "master" for reliefs.[1]

His aching ambition, however, was to paint. For this he showed a talent as real and raw as the rough stones he reworked, although like them he needed cutting, polishing, and setting. So in 1887 (or 1888) he enrolled in the Drawing School of the Art Promotion Society at St. Petersburg. He was desperately poor. He studied diligently and exhibited successfully in Copenhagen and Paris, but lacked the funds to do the task properly. His mother and sisters, equally famished, pleaded with him to send money. Stomach warred with mind, which warred with heart. At one point he is said to have contemplated suicide. In the end there was nothing for it but to return to Yekaterinburg, and make the Urals the subject of his art. These circumstances, unsought and unwanted, he turned to advantage.

He perceived the landscape differently from his more severely educated contemporaries. In particular he saw the Urals' forests aflame. For anyone committed to representing its indigenous scenery the topic would seem obvious: the boreal forest is a fire forest, its tempo of burning likely quickened in the late 19th century by Russia's economic liberalization and industrialization that broadened logging and threw sparks widely. Yet fire scenes were not a topic of academic interest, were not among the classics that students copied, were not a theme of Beaux arts. Those who painted fires typically came from the ranks of the untutored—naïve recorders of the world as they actually saw it, not a world learned by imitation from Old Masters. They painted fire in defiance of official indifference because it was so vivid and prominent. Fire and Aleksei Kuz'mich thus found common cause: the one proposing a distinctive subject, and the other, bringing sufficient skill to render that topic into formal painting. There was a long tradition in Russian art of fire icons, of Elijah the prophet on his fiery chariot, an image cherished to protect dwellings from lightning and flame. But giant canvasses of burning woods bore no relation to tiny portraits of Elijah. Flames ripping through the Urals required a different imagination.

From his youth he had been enthralled with nature's wonders. In that he differed little from most painterly contemporaries in an era aglow with landscapes. By temperament and circumstance he found himself among the amorphous second-generation school of Russian artists known as Wanderers, committed to folk themes, portraits, and especially natural scenery, and what evolved regionally into a Urals version of America's Hudson River School. He committed to explore the mountains and record its scenes. What drove him to wrestle with flames was apparently a fascination, which began to haunt him, regarding "grand fires" that he had witnessed in his youth. Reportedly, he suffered nightmares about conflagrations from which he woke up suffocating and coughing; he worried that he might hallucinate outright about fires; and to check the growing obsession he determined "to become one with fire" by fighting against flames near Shartash Lake. Still, he doubted whether anyone could truly capture the grandeur and power of a free-burning fire in its full-throated roar.

Yet he tried, over and over, seeking to distill fire's essence through countless sketches and fully wrought paintings. He began with a study of a grass fire (*Burning Grass*) in 1887, then escalated into the woods with a series of paintings all (confusingly) titled *Lesnoi pozhar*. Some were details, some were immense panoramas too vast to hang on walls (one actually looks down on a full-blown crown fire). Most disappeared, often mysteriously. A wonderful story, apparently apocryphal, tells how the 1897 version vanished from Perm during the civil war until, in 1934, a Red Army detachment carried the enormous canvas into town, having reputedly hauled it with them for over a decade. An 1898–99 version disappeared after being exhibited in Moscow.

His real breakthrough, however, came with an exhibit, *The Urals in Art*, staged on December 26, 1900, in Perm. It was an age of impressionism, when artists were experimenting with light, and Aleksei Kuz'mich added to the effect of his latest *Lesnoi pozhar* by displaying it in a single room with a solitary bulb. But exhibition lighting had less to do with the painting's appeal than his final mastery of a flame torching upward through the canopy. His earlier experiments had shown such flames running with the wind; in this quintessential version, it backs against the wind even as it ascends, causing a slow swirl that fixes the painting exactly in two and seems to draw all into a visual vortex. Aleksei subsequently carried his entire exhibit to St. Petersburg where it enjoyed genuine success, with *Lesnoi pozhar* as a centerpiece. Soon afterwards he added *Uralsky* to Denisov as a nom de plume; in 1903 he helped found the Siberian Society of Wanderers. The climax to over 15 years of studied fire painting arrived when he sent the work—the synthesis of all his efforts to render flame into art—to the 1904 Louisiana Purchase Exposition (the St. Louis World's Fair), where it walked away with a "big silver medal."[2]

More triumphs followed, happily for friends who considered Aleksei a sociable personality, much given to good humor. By 1912 commercial success allowed him to establish his own store, primarily for semiprecious stones and jewelry, across the street from Fabergé's in St. Petersburg; the imperial family was a frequent client. He donated funds to promote art in the Urals, sought to preserve the traditional crafts of ethnic groups in the region, and urged nature protection. His example inspired a virtual school of fire art through the works of L. N. Zukov, A. A. Sherementjev, N. M. Gushin, and I. I. Klimov, an organized corpus characteristic of no other country. But like that spiral of flame in his most famous painting, his moment of triumph was the outcome of peculiar, fleeting circumstances, and it was one that could not sustain itself.

Nemesis followed. He heard nothing, absolutely nothing, regarding the paintings he had sent for exhibition in St. Louis. (He was not alone; none of the artists knew what became of their works, and all were still protesting loudly at the Congress of Russian Artists when it convened in St. Petersburg in December 1911.) But the missing paintings were only a start. In short order both his mother and his son Nikolai died, and the Bolshevik revolution caught him in his Usekirka dacha, now within a newly independent Finland. He was isolated, geographically and emotionally. He felt "lost, forgotten, buried alive," one of the hordes of intellectual refugees the 20th century spawned like flies.[3]

He labored from his lonely retreat to reestablish contact with Yekaterinburg—to found a museum of art, to arrange to donate some 400 of his paintings, to promote his own repatriation. He failed. He remained marooned in Finland; his immense cache of paintings dissipated, and most vanished; even the iconic *Lesnoi pozhar*, to his knowledge, had disappeared without a trace. Queries regarding it went unanswered. Still in exile, his legacy apparently blown away like smoke, Aleksei Kuz'mich Denisov-Uralsky died in 1926.

For decades afterwards rumors swirled in Russian historiography about the ultimate fate of *Lesnoi pozhar* and the other contributions to the 1904 exposition, all associated with shenanigans in America. It was destroyed by a ship fire while sailing to an American buyer; a fraudulent art dealer in St. Louis had exhibited the work and never returned it; it was sold at auction to a mysterious collector, removed to Canada, and then to Argentina, or otherwise dispersed, the paintings lost, the artists uncompensated. The entire exhibit, it seemed, had, incredibly, vanished.

The true story exceeds those rumors: it is a sordid tale of war, politics, and fraud. It begins with efforts to entice Russia to participate in the exposition, which Russia did by arranging for a massive display of art, including some 600 paintings and a pavilion to hold them that one critic likened to a "cross between St. Basil's cathedral and a Hanseatic League warehouse," shipped in sections for assembly on site in January 1904. Within a month Japan and Russia were at war, and sensing that popular American sentiment was hostile, Russia officially withdrew from the exposition and shortly afterwards dismantled the pavilion.[4]

At this point, a Russian fur merchant, councilor of commerce for the Ministry of Finance, and member of the Russian Aid Organization Committee for the Exposition, Edward Mikhailovich Grunwaldt, stepped into the breach, having previously assisted during a troubled French exhibit in Moscow in 1891. Officially, Russia had withdrawn from St. Louis; but Grunwaldt was allowed to continue its program as a private individual. He subsequently arranged contracts with each of the contributing artists in which he guaranteed that he would either sell their wares or return them at his expense; if sold, the artists would get 70 percent of the revenue. He invested some $50,000 of his own wealth into the project and hastily pleaded for exhibit space. The collection arrived piecemeal and late and had to be housed on the second floor of the Central Arts Palace, all amid a relentless background of Japanese victories and American hostility. (Russia's abandoned exhibit space had gone to Japan, which enlarged its own contributions, thus recapitulating its military triumphs.)[5]

Still, the exhibition was massive and impressive, and among three special displays was one on *The Urals and Its Riches*. This was essentially Denisov-Uralsky's old exhibit updated, a reconnaissance of the Urals from "the scientific, geographical, ethnological, geological, mineralogical, petrographical, and artistic points of view" on what was characterized as "Russian California" by its premier "Painter-Mineralogist." There were studies of mountains, rivers, railway bridges, typical strata and minerals, mines and forests, all of which culminated in his masterpiece, *Lesnoi pozhar*. Such extensive burns were "the scourge of the Urals," a threat to its metallurgical industries, and a scene begging artistic description. All in all, *The Urals and Its Riches* constituted a tenth of the Russian exhibit. But the operatic fire painting, like a vast wild campfire, drew the greatest viewers and was praised as "most realistic and pictorial."[6]

That aesthetic directness was its salvation. *Lesnoi pozhar* was reproduced as a Sunday supplement by the Cleveland *Leader* on December 4, 1904—this with permission from the French Commissioner of Fine Arts, under the auspices of the French exhibition. (How the French muscled into granting permissions is unclear, but probably connects to Grunwaldt's previous experience with the 1891 French art exhibit in Moscow and his fur-traffic business in Paris, where his brother had an office, as did two of his partners in arranging the Russian collection.) Even with French promotion, however, the Russian paintings did not result in either the anticipated major prizes or big sales. An inquiry followed from the Lumber Insurers' General Agency to reproduce framed copies for distribution at national meetings—art to the service of advertising and fire prevention. But the Cleveland *Leader* was not the critical acclaim Grunwaldt craved, nor promotion by the Lumber Insurers the kind of monetary return he sought.

Disappointed, Grunwaldt arranged to try again, this time with official support, by shipping the entire exhibition to the Lewis and Clark Centennial Exposition in Portland, Oregon. When that scheme fell through, he decided to host his own exposition and shipped the entire exhibit to New York, a better art market and one for which he would not have to pay commissions to the exposition. In September 1905, with mediation from President Theodore Roosevelt, Japan and Russia ended their war by signing the Treaty of Portsmouth, while at a gala New York inaugural, attended even by Russia's ambassador to the United States, Grunwaldt opened *Russia's First Fine Arts Exposition in America*. In effect, he tried to achieve in New York what he had hoped would happen in St. Louis. Exhibiting presented no difficulties—Grunwaldt had wisely secured advance permission from the customs office. But selling the art did, and set into motion over the next eight years a prolonged legal battle in which fraud competed with fraud.[7]

Who owned the paintings? Grunwaldt implicated they were his legal property, although he tried to evade customs duties (they could be exhibited without charge but only if not sold). On March 7 and 8, 1906, he managed to sell 137 works at prices far below expected values to private collectors and institutions. Then on March 10 the U.S. treasury secretary intervened on behalf of both the Russian government and customs to end further sales and confiscated the lot as "unclaimed merchandise" until Grunwaldt paid the full import duties (those items already sold paid their tariffs, including eight destined for the Toledo Art Museum).

Grunwaldt could not meet the full charge, and in truth faced financial ruin. He owed $50,000 and had invested another $75,000 of his own money (in 2018 dollars, a tidy $3.2 million). By May Grunwaldt's lawyer, Henry I. Kowalsky, a probate attorney from California, reported that he was destitute, living in a basement, and suffering ill health. Grunwaldt appealed to the tsar for help, while Kowalski attributed his client's plight to American hostility toward Russia and to an incompetent auctioneer. (In August, Grunwaldt sued the Fifth Avenue Auction Rooms for $53,206, claiming criminal incompetence. The appeals lasted until June 1909, when they were finally dismissed.)[8]

Grunwaldt then prepared to return to Europe to confer with his brother, who ran their Paris office. Before departing he agreed, for the sum of one dollar, to "assign, sell, transfer and deliver" the full collection to Kowalsky apparently in the belief that the well-connected and politicized lawyer and lobbyist could break the logjam at Treasury. But Kowalsky, in the words of Robert Williams, could "only be described as a professional rogue." Among his other clients was King Leopold II of Belgium, until even that villain could stomach him no longer.[9]

When the warehouse bonds neared expiration, the Treasury Department prepared to sell the remaining paintings at auction. By now, the summer of 1907, the Russian government insisted that it was the legal owner, but declined to furnish an indemnity bond for the expected lawsuits. So although Russia claimed it owned the paintings, it refused to take steps to assert its rights; Grunwaldt, returning from Europe, insisted the collection was his, and hired another law firm; Kowalsky ostentatiously waved his contract with Grunwaldt to assert his ownership; and the Treasury Department maintained possession, with a threat of public auction, until the import duties were paid. Kowalsky borrowed enough to arrange an indemnity bond with customs and shipped the lot to Toronto where, in transit, ownership surreptitiously changed from Grunwaldt to Kowalsky. The money for this legerdemain Kowalsky had obtained by a personal loan, which he could not pay back, so he further wangled with another operator, Frank Havens, a developer in Oakland, California, who was erecting an art museum. With a prospective tariff war between the United States and Canada looming, Kowalksy and Havens moved quickly in March 1910 to ship the collection to Oakland.

Now the sad cycle repeated: Kowalsky claimed ownership, Havens held the bill of lading, and since the paintings were reentering the United States, the Treasury Department again demanded duties. The issue lay in limbo, partly paralyzed by the politics of the Taft Administration, until February 1912, when Treasury sold the collection at public auction. The ownership question devolved into money: the only contestant with enough was Frank Havens. He purchased the entire exhibit for $39,000. But Kowalski sued and pulled strings, and with Treasury still holding the paintings in limbo until the courts and politicians decided, the matter eventually went to the desk of President Taft in April 1912. He agreed that the auction should proceed. The Russian Fine Arts Exhibition for the 1904 Louisiana Purchase Exposition became the property of Frank Havens.[10]

As Robert Williams has aptly summarized this dismal saga,

The St. Louis Russian paintings disappeared into scattered private and museum collections in America after 1912 because of a series of ignorant and fraudulent acts by several individuals. In March 1906 Grunwaldt certainly tried to deceive both the artists in Russia and the American government by selling the paintings at auction without paying the tariff; artists had contracted with him to sell only certain paintings, and these at generally higher prices. The Russian ambassador tried to recover the paintings, made protests, but was unwilling in 1907 to pay the small charges and take the legal steps necessary to obtain them from the United States government. At this point two frauds were perpetrated upon both Grunwaldt and the Russian artists: Kowalsky shipped the paintings to Canada and took title away from Grunwaldt; and Frank Havens obtained the paintings from Kowalsky by purchasing them from the government at public auction, rather than accepting delivery of them on terms that would have included Kowalsky in the profits.

Kowalsky died shortly after the auction; Grunwaldt was ruined and died in 1915; Havens fell into financial distress and had to sell his collection at auction in 1916 before dying in 1918. The Bolshevik revolution severed any lingering involvement by the Russian government. In the end, the Russian artists got nothing, and the paintings, other than those sold in New York, were blown to the four winds.[11]

And *Lesnoi pozhar*? So far as the Russians knew, it was among the disappeared. Grunwaldt implied in letters that he had all the paintings in his possession when he arranged the exhibit and auction in New York over the winter of 1905–6. But *Lesnoi pozhar*, now identified by its English title *The Forest Fire*, was not listed among the confiscated merchandise, and was not among pieces registered for the auctions of 1907 and 1912. In fact, it had never left St. Louis. Shortly after it was reproduced for the Cleveland *Leader*, Adolphus Busch bought it and hung it in his mansion.

Busch was a likely buyer. He was on the Exposition's Committee of Fifty, one of the largest subscribers to exposition stock, a member of the directory, and chairman of the Committee on Foreign Relations. He and Grunwaldt would have known each other, the Russian exhibition surely had Busch's support, and Grunwaldt would have been desperate for sales and perhaps hoped Busch might arrange for others on the committee to help. For Adolphus Busch the painting was a majestic souvenir of a successful enterprise; for Grunwaldt, a hopeful trophy sale, perhaps discounted in gratitude and expectation.[12]

The transaction had a commercial logic. After all the Louisiana Purchase Exposition, rather in keeping with its origins in a century-old land buy, had from its conception been a business proposition with little regard for those on the scene; now, *The Forest Fire* had also passed hands, businessman to businessman, with little say from its creator. But the sale had its esthetic logic as well. *The Forest Fire* was not among the paintings fawned over by professional critics or buyers; its artist was still a relative unknown outside certain Russian circles, and its subject completely outside the motifs of academic and professional art. It was the kind of painting that, appealing to popular tastes, would be reproduced in Sunday supplements, handed out as advertising by insurance companies and later forestry associations, and purchased by a captain of industry who liked its gargantuan size and easily understood theme. It would get copied and capture public imagination, and so was spared the dismal exile of the Russian exhibition generally.[13]

Meanwhile, civic leaders in Dallas, Texas, enticed Adolphus Busch to erect a grand hotel commensurate with their municipal ambitions. The result was a Beaux arts edifice, known generally as the Adolphus, that opened in October 1912 and was for many years the tallest building in Texas. When the Adolphus was refurbished in 1926, *The Forest Fire* went to the foyer to add a touch of opulence, a kind of exalted if wild European grandeur, popular taste with an Old World cast and a heroic scale in keeping with its announced ambitions. (Its size and realism was such that it was said that one "must not stand too close because of the heat.") The painting remained until the next refurbishing in 1950, when it was relocated to the Anheuser-Busch Brewery in St. Louis. There it presided over the "Hospitality Room" and was subsequently viewed by tens of thousands of visitors through a haze of beer and cigarette smoke, an exposure far greater than any it might have enjoyed in the Havens Gallery or the Toledo Art Museum.[14]

Even before it graced the Adolphus Hotel, it had become the best known of the Russian entries. The 1904 reproduction by American Colortype was joined by a 1909 edition. The lithograph was subsequently broadcast widely by the A. G. Voss company under the title *The Untamed*

Element. The Chicago *Sunday Examiner and America* promoted "exact reproductions" as a promotional feature for subscribers. The U.S. Forest Service included a black-and-white photo of this edition, somewhat cropped, in its historic photo collection and identified its source as A. G. Voss, May 18, 1949 (although the date may be a simple typo for 1909). The American Forest Association advertised reproductions as a *Sermon in Color* for its fire prevention campaigns. The painting propagated throughout rural America with an abandon entirely appropriate to its subject. Through its several published versions, *The Untamed Element* got into homes and work places. There, it has been copied again, and again, and again. There is no telling how many amateur artists have practiced their craft by reproducing the scene in their own idiom. Not a few have been cheeky enough to pass the work off as their own composition, or to suggest that their version in oils or acrylics might be the original.[15]

I have myself accidentally added to this tale of proliferation. While we were doing the raw research for *Fire in America*, my wife, Sonja, found a print at a secondhand store in Altoona, Pennsylvania. Then, Princeton University Press and I had difficulties agreeing on either a title or a jacket cover that we both liked. (The book was actually advertised prior to publication with a different title until my editor demanded the shortest possible version. I couldn't think of anything briefer than *Fire in America*.) And I wanted an image of a fire on the cover. At one point I proposed a Currier and Ives print of a prairie fire, but Princeton disapproved, so I argued for a wonderful forest fire painting I had collected. To spike interest I sent the black-and-white print from the Forest Service collection, which was eventually used for the inside title page.

But I insisted on a color version for the cover, so I hurriedly photographed my lithograph with 35 mm slide film and sent the image to the press to show the possibilities, expecting that I would ship a print later. I didn't hear back, and then found they had reproduced the cover from the slide, as is, although with the image reversed. Unfortunately I had shot the slide indoors with an incandescent bulb that cast a deep yellowish hue over the aging lithograph, rendering it both more vibrant and more sinister than the original. The paperback edition later issued by the University of Washington Press, by focusing even more closely on that flawed image, distorted it further from the original. Only a handful of people would recognize in the book's cover that archetypal image from Denisov-Uralsky. So the saga continues.

And the original? An American historian of Russia, Robert C. Williams, spotted the painting at the Anheuser-Busch brewery in 1971, became intrigued over its provenance, and tracked down the sordid story of the Russian paintings at the exposition. An article he wrote attracted the attention of former Missouri Congressman James Symington, then a lawyer in Washington, D.C., who urged August Busch Jr. to donate the painting to the U.S. government, who could in turn transfer it to the Soviets. Busch agreed, and in March 1979 Joseph Duffey, chairman of the National Endowment for the Humanities, handed over *Lesnoi pozhar* with all due ceremony and a round of vodka toasts at the Soviet embassy.

Still, Busch's was an odd gesture at an awkward time. The U.S. bicentennial had long ended, and détente had collapsed with the Soviet invasion of Afghanistan in 1978. Why would a hardnosed businessman like "Gussie" Busch donate what could legitimately be reckoned a family heirloom? He offered no exegesis, though a likely explanation may be that he hoped it would help land the beer concession for the 1980 Moscow Olympics. If so, that ambition lost its fizz when the United States subsequently boycotted the event. Once again, the painting had become a pawn to an unstable alloy of entrepreneurialism, war, and politics.[16]

At least it seemed that the out-of-Siberian exile of *Lesnoi pozhar* was ending. Ambassador Anatoly Dobrynin joked, during the transfer ceremony, about keeping the painting in the embassy but worried about how a fire painting might be interpreted, then suggested it would go to an appropriate museum, perhaps Tretyakov or Perm, he wasn't sure. A return to the Urals made sense; over the years at least some of Denisov-Uralsky's art had wended its way back to the mountains and forests that had inspired it, installed as special collections within the art museums at Yekaterinburg and Perm. Upon learning of the putative repatriation of *Lesnoi pozhar*, the curator of 18th- to 20th-century Russian art at the Perm State Art Gallery requested of the Ministry of Culture (USSR) that the painting come to its galleries, where it might join the others, including the 1897 version. There was no reply. The painting had again vanished.[17]

———

Where it hangs now is unknown, or whether it has again been pilfered or exploited, or if it even resides in Russia. Like its flames, the physical painting seems to dance from spot to spot, flaring and smoldering against the winds—a cultural vortex collecting and consuming the quotidian world and lofting it to a kind of transcendence, yet doing so against the perversions and wiles and greed of politics and business. It captures a dazzling instant of struggle, its compelling visual tension deriving from the thrust of its aspirational flame against the ambient world. Whether Aleksei Kuz'mich was satisfied that he had caught the essence of fire is unknown, but he certainly made vivid a moment of history and the enduring tension to make art out of a force of nature.

Whatever the original inspiration, the painting has imprinted itself on cross-cultural imaginations, and whatever the fate of the originating *Lesnoi pozhar*, its reproductions seem deathless. In its history no less than its imagery *The Untamed Element* offers a suite of distilled reversals from the expected. It shows a fire flung upward to rather than descended from the heavens. It describes an exile out of rather than into Siberia. It offers an original than may, in the end, be less than its imitations. It displays an archetypal image almost Jungian in its instincts, a pillar of fire that testifies not to a Deity but to a deified Nature, a secular icon more widely known than any of Elijah in his fiery chariot, although unlike those of the prophet it could not promise to shield its creator from harm.

That image continues to percolate through the American public. Reproductions still decorate walls, fill second-hand stores, and inspire local artistes. And as for those recherché original oils on eBay—caveat emptor.

———

Postscript: The Yekaterinburg Art Museum hosted a 150th celebration exhibit of Denisov-Uralsky's birth with a major exhibit of all his various arts. It was unable, however, to locate *Lesnoi pozhar*. Then, while the exhibit was closing, word came that it might have been located in the basement of a museum in Tomsk. A photograph and its dimensions suggest it is in fact the fugitive painting. As yet no one has positively identified it nor restored it, but its long odyssey homeward may finally be ending.[18]

FIRE FAR

W E DON'T KNOW THINGS in themselves. We know them in context. We know them by comparison, by contrast, by history. Much of the American fire scene shares a similar narrative and pyrogeography with other parts of the world; much does not. It's useful to see where we align and where American exceptionalism might be at work. "Fire Far" collects some examples from other fire powers. It ends with a land without fire.

EAST OF THE WIND AND
WEST OF THE RAIN

I had intended to join the 20th anniversary celebration of the great fire experiment at Bor Island, but Russia denied me a visa. Instead I found myself crafting a proxy tribute to Johann Georg Goldammer, one of the most extraordinary fire personalities over the past 40 years of fire on Earth.

THERE ARE PLACES THAT REST tangibly on the Earth's surface, and places that flourish only in the imagination, and places that site their existence within a moral geography, and a few places, not many, Bor Island among them, that manage to fuse all these settings together. In truth, Bor belongs with that long tradition of island arcadias that have attracted Western thinkers since well before Thomas More in 1516 gave them the name they now have: Utopia. What makes Bor Island unique is that its informing theme is fire.

The tangible Bor Island is a 50 ha patch of sand and Scots pine along the east bank of the Yenisey River halfway between Krasnoyarsk and the Arctic Ocean. It sits close to the geographic center of Russia. Its imaginative existence appeared in 1992 when Johann Georg Goldammer, a German forester turned ecologist, saw its potential as a site for an international experiment and named it after a small village, Bor (Russian for "conifer forest"), a few kilometers to the north. It is a contained space, surrounded by marsh and river, an ideal setting for an experiment, but especially for one with fire. It metamorphosed into a moral landscape when it passed through the portal of Goldammer's idealism. The international cadre of scientists who conducted the burning came from countries that two years before had been Cold War rivals, and this made Bor Island an experiment in political ecology as well. Its immolating fire could be imagined as a kind of Ragnarok, the world-ending blaze of Norse mythology, intended to burn away the legacies of the old regime. More broadly, the conifer-clad isle was a quirky microcosm of Earth, the only planet known to have fire, and of how its dominant, uniquely fire-using creature might inhabit it. Our species' monopoly over fire has granted people a defining ecological power, but as we moved from burning savannas to burning fossil fuels, we unhinged the dynamics of the planet.

Johann Goldammer's vision was to burn off the isle, measure the character of the fire, sample its emissions, and then, over the next 200 years, record what regrew in its aftermath. It was an improbable quest—better suited for a fairy tale—from someone only halfway through an improbable career.

I first met Johann in 1987 at a fire and meteorology conference in San Diego. I was coming off a big-screen history of fire in America and a textbook on wildland fire, and I had decided the best way to learn more was to go broadly comparative rather than drill more deeply into what I had; I had spent a season in Antarctica and had completed a long research tour of Australia. Johann was enduring his *habilitation*, a probation period in German academics, to become a professor of forestry at Freiburg in the course of which he introduced the term *Feuerökologie* to German and the concept to German forestry and had studied fire in Brazil and India. The international fire community was only beginning to consider global issues, so the conference was overrun with Americans. Johann slouched on a sofa, watching the spectacle but with a mind clearly elsewhere. If he seemed out of place, it was likely because German forestry had almost no tradition of fire studies and American fire conferences didn't embrace much of the world outside North America. We chatted and both agreed that fire belonged to the Earth. Out of passion and necessity, he was already the fire community's first cosmopolitan.

We stayed in contact and renewed our friendship at a conference on fire in the tropics that he organized in Freiburg the following year. In 1991, when I was planning a month-long fire study tour of the Soviet Union,

I invited him to join our party. This would be the first American contact with the Soviet fire community since the 1978 invasion of Afghanistan and probably the first German contact ever. Two Swedes, also invited, subsequently dropped out. Johann and I went together, hosted by Avialesookhrana, the country's aerial fire protection service. Our first night, at the Intourist hotel in Leningrad, he told me his story.

———

He had grown up in Marburg, close to the center of Germany. His mother, Inge, was German; his father, Kurt, had a Polish mother. Nazi race laws denied them a marriage license. Then, on March 28, 1945, American troops marched into Marburg. His father immediately revisited the mayor and demanded a license. The mayor again refused. "You know the law." Kurt Goldammer seized the mayor by the lapels and spoke very slowly into his face. "I don't think you fully understand what has happened today." They were married the next day. Johann was born on August 23, 1949. The Goldammers had a soft spot for America ever after.

He was his parents' child. From his father, a professor of religious studies at the University of Marburg and the long-serving president of the Paracelsus Society, he learned to value scholarship and the Paracelsian notion that toxicity resides in the dosage not the substance. From his mother's family, with long ties to the North Sea, he caught the "seafarer virus," which led him to work on a cargo ship the summer he turned fifteen and, from 1968 to 1972, to attend the Naval Academy Flensburg-Mürwik. He graduated as a *Leutnant zur See*, was later promoted to *Fregattenkapitän*, and remained in the naval reserve as the commander of a minesweepers and mine hunters. Whether that military experience only sharpened his talents or helped instill them, he has an exceptional capacity to organize, a quiet authority, and the ability to command. In a sense his career on land has emulated his experience at sea. He is the person who clears away the impediments and makes it possible for others to maneuver.

The prime mover of Johann's character, however, is a deep-bred idealism, and this was joined to a fascination with forests when, still in his teens, a supervisory forester gave him a tour of a private estate. The "forestry virus" joined the seafarer virus. In 1972 he enrolled for a diploma in forestry at Freiburg University. A casual observation from one of his professors, Jean-Pierre Vité, and his family's lingering sympathy for things

American pointed him in a direction far from the traditional themes of German forestry. The Americans, he learned, were deliberately burning some of their woods. Johann went to the Tall Timbers Research Station, north of Tallahassee, Florida, to find out for himself. There he was mentored "by the Nestor of fire ecology," Ed Komarek. "Fire Ecology and Fire Management" became the subject of his diploma thesis. In July 1977 he joined the State Forest Service of Hesse. After two years he returned to Freiburg for a doctorate in forest science, this time studying fire in southern Brazil. While in Parana, he married Dorothea Knappe, who, he proudly notes, became the first woman to conduct a prescribed burn in Brazil's pine plantations. He graduated in 1984. In landscape fire he found a theme common to all peoples and most places—and a medium for his idealism. Fire could integrate the disparate parts of his self as it does for the Earth's mountains, forests, and winds.

The contours of his career came into focus the next summer. A Greek charitable foundation invited him to discuss with Greek colleagues the fires then plaguing that country. On August 18, 1985, while boating to Thassos Island, he could see a smoke plume swelling upward. At Limenaria he watched the Greek navy readying to attack the flames and offered his services as both a naval officer and a civilian fire expert. He was accepted and promptly organized a gang of sailors and "instructed them how to fight the fires" with the buckets, hand tools, and towels on hand, and then he worked with villagers from Maries. At that moment he discovered what he described to me as "the two different souls in the two chambers of my heart." One belonged to a "forester and ecologist," the other to "the Captain of a ship." At Thassos he first "felt the unity, the symbiosis, of both."

Two years later we met in San Diego. In July 1991, we converged on Helsinki before flying to Leningrad, then full of glasnost, perestroika, and a Soviet fire community eager to connect with the West. Johann interested them as a representative of European traditions—Germany had long been the portal for Russian forestry. I interested them not as a historian (my academic posting) but as a fire specialist and the author of *Introduction to Wildland Fire*. Apart from a few days at the Leningrad Institute of Forestry and Avialesookhrana's Pushkino headquarters, outside Moscow, we toured east of the Urals. Johann had to depart from Yakutsk to attend planning meetings for a large fire experiment in Southeast Asia. I completed the traverse to Khabarovsk and the fire-and-tiger-rich Sikhote-Alin' Mountains. The coup attempt that led to the dissolution of the Soviet Union occurred a few days after my return to the United States in early August.

Roughly midway in our tour, at Krasnoyarsk, where both the Forestry Ministry and the Soviet Academy of Sciences had fire labs, Johann conceived the idea of an international conference and a large-scale field experiment that could unite former rivals in a common cause. The next year I hosted a delegation of Russians in the United States and Johann found funding from the Volkswagen Foundation and identified Bor Island as an experimental site. In 1993, the Sukachev Institute of Forestry hosted the conference in Krasnoyarsk, and then Johann led the group to Bor Island.

By then, however, our paths in the flaming woods were diverging. I took an academic track and wrote books. Johann took a more entrepreneurial track and invented institutions. I participated in the Krasnoyarsk conference and then went east to Buryatia and Chita, the scene of mammoth fires that had occurred in 1987, known to the West only through declassified satellite imagery. Johann burned Bor.

On July 6, 1993, the Bor Island group—Finns, Canadians, Americans, Chinese, Swedes, and Russians—under his leadership and with operational assistance from Avialesookhrana ringed the island with fire, sending an enormous plume upward in what resembled a giant burnt offering. Many of the team returned the next year to resample their transects. They claimed the beetles, gorging on the blasted pine, were so loud they drowned out the ceaseless hum of mosquitoes and black flies. Most of the island, though not all, was burned clean by flames lofting from a dense lichen prairie through the conifer canopy. A worn-out world was about to be replaced by a new one growing through the ash.

———————————

What had been a career trajectory became a quest. We met at conferences—often those he had organized—and compared notes. We exchanged seasonal greetings, he guided me to some central European fire history sources, and he called me on the eve of German unification. "An important event," but also "a takeover by the West," he thought. But I had to watch from afar as he created a unique niche as a fire impresario. The closest historical analogy I could contrive was Dietrich Brandis, the botanist turned forester who created a forestry service for imperial Britain in the 19th century. Over the next thirty years there was no required skill too difficult for Johann to master, no task too arduous to undertake, no place too remote to visit. He completed his *habilitation* at Freiburg, which qualified him for a university professorship. He arranged conferences on fire's ecology and management and published their proceedings. He set about acquiring the practical skills he needed. He learned fire fighting and fire lighting and their tools. His English became more fluent, and he became conversant in French, Spanish, and Portuguese. He trained for a pilot's license to fly small planes. He was certified in explosives. He learned the evolving sciences that connected fire with the atmosphere and satellite reconnaissance and so earned an appointment with the Max Planck Institute for Chemistry under the leadership of Nobel laureate Paul Crutzen and Meinrat O. Andreae.

He became the Earth's most widely traveled fire expert. On behalf of the United Nations and international development organizations, he went to Brazil, the Philippines, Myanmar, India, and Indonesia. He journeyed from Mozambique to Mongolia, from Albania to Nepal, from Ethiopia to El Salvador. Everywhere he sought to build capacity by organizing conferences, editing books, and transferring expertise. He helped plan complex, multinational field campaigns to measure fire and its effects in southern Africa, Brazil, Southeast Asia, and Eurasia. It was not enough to do science or write up a consulting tour: what mattered was creating institutions that could endure, cope with fire on the ground, and continue the job.

A career like this has so much momentum that it would simply foam over normal obstacles and, if checked in one channel, splash over into another. It would have happened even without our visit to the USSR. But that experience alerted Johann to the empire of fire to the east, fashioning a boreal counterweight to his passion for the subtropics, and the implosion of the Soviet Union created opportunities for someone of his temperament because those former states became vacuums of fire administration, ready for advice and guidance. And there was Russia itself, with

Bor Island close to its geographic core, much as the taiga was central to its psyche. Over the next 20 years, the former USSR and its satellite countries would claim Johann's special attention, with Bor Island as a geodetic marker for how he thought about fire's ecology, about pyric politics, and about his own standing in what he was coming to call the "fire globe." Bor was the place that, by choice, he returned to.

His unlikely career as a fire authority—implausible not least because Germany had no tradition of fire research or institutions beyond simple firefighting—seemed to rise like that plume over Bor. In 1993 he became the leader of the UNECE/FAO Team of Specialists on Forest Fire and edited its *International Forest Fire News*. In 1998 he created the Global Fire Monitoring Center (GFMC), located in the abandoned control tower at the Freiburg airport, built of steel and straw mats in the 1920s and topped with a windsock, rising out of the tarmac like a castle tower. The GFMC quickly became the bridgehead for a planet-spanning consortium of regional fire programs. It was a nongovernmental organization for global fire, until absorbed into the UN International Strategy for Disaster Reduction (ISDR). Johann had campaigned tirelessly for the ISDR to add fire to its roster of earthquakes, volcanoes, floods, and hurricanes. The UN rubric brought a measure of credibility and funding.

Many of the world's fire hotspots were also political flashpoints, and he was there to find common cause in better fire management. Unsurprisingly, perhaps, a good fraction were former Soviet provinces or client states. Ukraine, Georgia, the Baltics, the Balkans, Azerbaijan and Armenia, and Mongolia—he offered conferences, confidence-building missions, and training exercises. A man who seemed never demoralized, he was always there to encourage. Around a common fire, emblematic of our common humanity, it was always possible to hope.

When a reporter with the *Wall Street Journal* called me in 1999 to help write a story about issues with global fire, I told him he should talk to Johann instead, and he called back to say I was right. I continued to write, expecting that somehow my words would lead to or at least shape actions, but Johann acted, and then recorded in words. There is a difference between keeping the flame and carrying it. The numberless hours sitting on folding chairs at droning conferences, the countless clichéd toasts, the jumbled millions of polyglot words—he has patiently endured them all and tweaked them into declarations, position papers, manuals, and ultimately institutions that could improve fire's management, and, through fire, humanity's stewardship of the planet.

The GFMC website posts his personal calendar which, for practical purposes, is indistinguishable from his professional schedule. It records a blistering pace: just reading the relentless engagements, week after week, continent after continent, would exhaust a normal jet setter. My books might fill shelves, but his workshops put fire boots on the ground. I had spent eighteen fire seasons in the field with the National Park Service. I knew what mattered most. I dedicated *Vestal Fire*, my fire history of Europe, which included a long text on Russia, to him. *Mit vielen Grüßen*. With much feeling.

But like all quests Johann's had its costs. Some were personal, and when we met at conferences he would confide a few to me. Funding sources sometimes faltered, losses he made up with his personal money. Official indifference, sometimes bordering on rudeness. The sense among academics and geopoliticians that fire did not much matter, that it was something that happened in outbacks and backcountries that had picturesque scenery and grizzly bears, and was addressed by forestry, an applied practice, not by real science, that big burns were a freaky sideshow in the planet's unfolding ecological drama, not an index of global change and maybe a contributing cause to it. And mostly, the absence of time for anything but the cause. His family knew him more from his absence than his presence, yet Dorothea remained as stalwart as when they had once gripped a driptorch together in Brazil. She and their daughter, Katerina, "patiently, sometimes desperately," as he put it, kept the home fires burning while he trekked to hotspots around the Earth.

A grail quest, yes. And like all quests, his furious pursuit of fire had not just been an adventure but a series of trials.

———

Among the major firepowers, Johann's position is curious. What holds his peculiar empire together is his personal vision. The more arrogant of the global fire community may politely regard him as a kind of Peter the Hermit on a children's crusade. He commands no battalion of engines, can launch no fleet of air tankers, oversees no big-science lab. He must forage constantly for funding. Fire scientists suspect the caliber of his research, so pointedly practical and often political. The fire Establishment may dismiss the Global Fire Monitoring Center as a kind of Vatican among the real powers. Under their breath they might mutter, after Stalin, How many divisions has the Pope? But if you add up all the networks, inter-connect all the collaborative projects, weigh the ardent students recruited to the cause, adjust the fulcrum for the scores of leveraging effects, and if, like Johann Goldammer, you can cajole and inspire and never be driven to despair, the answer might be, Quite a few. Because he does not represent a national government, he can move where the major powers cannot; he has succeeded where larger, less nimble, and less motivated organizations have stumbled.

And he has patience. In March 2012, after five years of persistent prod-ding, he finally received permission to burn off, for ecological regener-ation, a heath at Jüterbog Ost, a former Soviet-German artillery range, now converted into the Heidehof-Golmberg Nature Reserve. Five years is not too long for someone who considers 200 years a reasonable duration for a fire experiment.

———

Throughout, Bor Island continued to beckon. Over the 20 years following the original experimental burn, he visited Russia 42 times, 6 with side trips to measure the changes the grand experiment had wrought. Twice the Russian Forest Ministry awarded him medals; once, by invitation, he addressed the Duma concerning the revised but flawed forestry law it had passed. In July 2013, with mandatory retirement from the Max Planck Institute and Freiburg University approaching, he organized what would be his last trek to Bor, a 20th-anniversary reunion during which the future would be handed over to the next generation. I was to join him, a chance to rekindle our friendship in the field. The future, however, was out of our hands.

In the immediate aftermath of the Bor burn, the island had sprouted lush regrowth. Fire did what it so often does in nature and myth: it renewed. The contrast with the slow rot and riot of the Yeltsin years was striking. But part of the paradox of fire is that it is also conservative. Unless the fundamentals change—unless new species arrive or old ones disappear, unless the wind and rain alter their rhythms—the new growth will assume the form of the old. Revolutionary fires typically end with a reformed regime that, in its basics, resembles what it putatively replaced. And so it has proved at Bor Island, where the soft tyranny of Vladimir Putin shut down the special quality of ecumenical science that had characterized the original experiment. Except for Goldammer, the Russians denied visas, mine among them, to anyone not Russian or a member of a country that had once belonged to the Soviet Union. Johann was left alone for his final visit. A *Pinus silvestris* forest was regrowing much like the one burned away on that bright July day in 1993.

It might seem to most observers a sad coda to a quixotic dream, but Johann Goldammer knows the value of tenacity and endurance. He knows that the tree you plant one year will grow to shade another generation. Given time and a change in global climate—as well as the climate of opinion—the outcomes might shift. The work goes on. He hardly sleeps. There are meetings on capacity building, aerial fire control, and research on fire in the Earth system to organize and coordinate in China, Austria, Sumatra, Brussels, Italy, Mongolia, and Croatia. There are training exercises to conduct in Georgia. There is an international fire conference to organize for Korea in 2015. It can seem a frenzy of errantry.

Yet there remains a vision at the core, and it was once given a tangible form as Bor Island. If it seems an odd place, recall that folk tales abound of improbable youths called to unlikely roles and of treks to scenes strange beyond reckoning, and that intellectuals have repeatedly dreamed of arcadias idealized to the yearnings of their times. If you believe in the commonwealth of humanity, if you have adopted fire as the medium of your idealism, if you have talents for organization and command, and if you wish to live by your convictions, then you must unflinchingly follow your heart's desire to wherever the flame beckons, even if that quest takes you to a pine-clad sandbar that lies east of the wind and west of the rain.

STRABRECHTSE HEIDE

A meditation on the role of scale in fire ecology, and the place of fire in European land use and agronomy.

THE DUTCH ARE the tallest people in Europe, and Holland, Europe's most densely populated country. It is an odd pairing, but then the Dutch have long had more stature on the world stage than the size of their country suggests. Allowing for some exaggeration, their story is a distillation of Europe's. Certainly that is true for their landscapes.

The Strabrechtse Heide is a patch of deliberately rough country embedded within an intricate mosaic of rural, urban, and industrial tiles, squares and rectangles fringed with streets, hedgerows, and canals. Coming upon it almost startles, like finding a splash of spilled paint on a Piet Mondrian canvas. The Heide is a chimera of Calluna heath, grasses, wetlands, shrubs, and pine woods, one of the gems of Netherlands' nature conservation. It sprawls tidily over a mere 2,000 ha—a jewel box compared with big-box parks like Yellowstone, Kruger, or Wood Buffalo. (The largest of Dutch protected landscapes barely qualify for the smallest of American wilderness sites. Yellowstone has a land mass roughly a fifth that of the Netherlands.) That there is any fire at all is both inevitable and astonishing.

<hr>

The Netherlands—the Low Countries—are a delta for a multibranched Rhine, and are correctly known far more for water than for fire. But with allowances for its environmental peculiarities, the Netherlands are also a kind of historical delta through which flow the classic landscapes of temperate Europe. They are, in shorthand, a garden.

As it stabilized after the Middle Ages, that scene had three components—the arable field, the pasture, and the woods, or in the language of Latinate agronomy, the *ager*, *saltus*, and *silva*. These were cultural landscapes, as expressed literally in such terms as *agriculture* and *silviculture*. Each component featured fire in some fashion. Colonizers cleared with fire and axe, farmers burned arable land fallow according to regular rhythms, pastoralists fired rough patches that often blurred into arable and woods, and the woods knew fire through swidden cultivators, patchy burns to stimulate berries and mushrooms or to expose mast, and escapes from pastoralists. All this burning, however, occurred within an anthropogenic matrix. In temperate Europe, where seasons recorded a change in temperature rather than in precipitation, there was no routine cadence of wetting and drying, much less regular bouts of dry lightning. Like the landscape within which it resided, fire was an artifact of people. In particular it depended on the availability of fallow, which in a cultivated landscape had to be tolerated, if not grown outright. The fallowed lands also furnished most of the landscape biodiversity of habitats and species.

During the agrarian revolution that gathered strength during the 18th century, reformers, with the assistance of the state, sought to intensify production on each of the tripartite features of the classic landscape. This was particularly true for agriculture and silviculture, the latter evolving into forestry. With special vehemence theorists and ministries denounced both fire and fallow. They made explicit that what divided modern from primitive agriculture was that primitives used fire, while moderns found alternative means to fumigate, fertilize, and otherwise catalyze and recycle according to the principles that made this kind of farming an exercise in applied fire ecology. Accordingly, a long-simmering distrust of fire in the hands of peasants and herders mutated into a more virulent animus. If fire was a tool of cultivation, officials and academics demanded a different tool. After all, fire had no legitimate basis in nature. It existed because people put it in. They could also take it out.

―――――――

Within temperate Europe, the model morphed as it adapted to the oddities of local circumstances, substituting different kinds of trees, for example, or using coppice or shrubs. Then it pushed against those geographic frontiers. Sweden carried the model into the boreal forest, converting swathe after swathe of coniferous forest into swidden fields, and eventually into tree farms. France thrust the model, now remade by long tenure in temperate environments, back into its Mediterranean origin, and thereby commencing what became a ceaseless firefight from Provence to Algeria. Russia projected the model from the mixed-forest realm of central Europe into the coniferous taiga of Eurasia. Britain made the European anomaly the imperial norm, and so expansively exported the ideal that the sun never set on it.

And Holland? Holland enlarged the model, polder by polder, into the deepening off-shore of the Atlantic, reclaiming the sea for farms. In the 17th century it joined other seafaring countries and leaped over the Atlantic altogether to plant trading factories in the Americas, Africa, and Asia. The Dutch East India Company was not interested in settlement or conquest but in commerce. It in fact discouraged demographic colonization at Cape Town, and elsewhere; seized Brazil's sugar colonies for only a couple of decades; acquired a handful of tropical islands; and only really got sucked—reluctantly—into imperial conquest in the East Indies, where they became involved in plantation forests for teak. Such settings attracted fires like black rats to East Indiamen galleons.

Throughout, Dutch explorers and traders encountered fire as a normal landscape feature on a scale unimaginable not only in waterlogged Holland but over all of temperate Europe. Where they could, they condemned fire, and punished those who set them. But the dominant encounter of fire-intolerant Europe and a fire-prone Earth was left to the major imperialists, notably the British, French, Russians, and the Neo-Europeans birthed in settlement societies such as America, Canada, and Australia.

The clash was violent: the social collision between different peoples had its environmental counterpart, as one firefight mirrored another. The most spectacular expression followed the establishment of forest reserves that were turned over to professional foresters to administer. State-sponsored forestry created the apparatus of modern-era fire protection; its institutions, its science, its purposes. Yet forestry was a graft onto the great rootstock of European agriculture. It might, for a while, tolerate fire, but it sought, as an ultimate intention, fire's extirpation. Until fire was controlled, modern forestry was impossible.

This ambition set into motion a heroic experiment in fire exclusion whose full costs, both economic and ecological, have become increasingly apparent during the past few decades. Of course forestry was only a part of the attempt to reconstruct alien landscapes wholesale into something more familiar, more productive, and more secure. Controlling fire was a means not only to control the countryside but to control the peoples who resided on them. This political imperative was never far from the minds of those who decided in what ways fire ought to exist. Garden landscapes required garden-like societies in which everyone had his place and time, even if such a vision was as rare as those outlier European landscapes that knew no regular rhythm of wetting and drying.

This story of exported pyrophobia is now well known, and is recycled like phosphorus through the narrative ecology of environmental history. The fire community in the Big Four firepowers appreciates thoroughly the ruinous folly of imperial Europe trying to impose its vision of a fire-free world onto taiga, tallgrass prairie, savanna veld, and pine steppelands. But just as once-colonized nations tended to become, in turn, colonists, seemingly forgetful of all they found detestable and foolish about ignorant imperialists, so the new firepowers tend to promote their own understandings as normative. The absurdity of applying the ecology of Strabrechtse Heide to America's Yellowstone or Australia's Kosciusko national parks, however, is no less absurd than transferring the fire ecology of such places to the Heide.

Northern Europe's great heathlands sat awkwardly within the manicured landscape of field and forest. They were an accidental byproduct of clearing, burning, and grazing in which the woods failed to return, the land often became waterlogged, and the minor flora that had previously underlain the forest or had clung to impauperate soils amid sand and rock spread promiscuously. Grazing and burning kept them in woody heath, often pocked with grasses and wetlands. On poorer soils they were extensive; a survey in 1850 reckoned the Netherlands' heath cover at 750,000–800,000 ha. Mostly it was used as rough pasture, for which fire was mandatory to keep the browse palatable to sheep and cattle.

It was the kind of scene that drove academic agronomists and ministers of agriculture into a frenzy. Such landscapes were, in their lexicon, "waste." Worse, it was burned, which doubled the wastage. If browsed by sheep, fire and hoof worked to replace Calluna heath with the grasses that sheep preferred. States labored to convert the wastelands to arable, usually by paring and burning—that is, swidden applied to organic soils. Or they sought to plant it to trees. Or, as in Scotland, it became prime habitat for grouse and an economy of sport hunting. The heath itself had little value: even animal husbandry sought to improve itself through breeding, close tending, and integrating the flock better with the field. Still, pastoral burning persisted through the 1920s on ever-shrinking parcels of heath.

Everywhere, the rolling heathlands shriveled, and might have vanished altogether in the face of woody reclamation and reduced grazing had not something unexpectedly intervened. In many sites, vandals burned the heath for sport—and paradoxically spared it from destruction. Other sites became the scene for military camps; exploding shells took the place of paring and burning, and the tread of tanks substituted for the hoof and teeth of sheep, cattle, and swine. And here and there, as the heath threatened to disappear, champions arose to declare them cultural relicts, worthy of at least ceremonial preservation.

That is what happened to Strabrechtse Heide. In 1950 the Netherlands state forestry bureau bought 1,000 ha as a preserve, an act that so infuriated the local Boers that between 1952 and 1954 they burned off almost the entire site in protest, an outburst that may have saved the Heide from biotic decay. Within a decade the foresters reinstated some sheep, and later some highland cattle, all under close tending, and they burned to promote grass, which the sheep favored, over the Calluna that ecologically defined the heath. In the 1970s the local authorities banned open burning altogether. The Heide overseers ignored the ban, and continued to fire the site, even though in patches little bigger than carports and with less depth than a strong rake. Something had to prune the woody growth. Hand clearing was too expensive; mechanical treatments often too crude; it was either fire and livestock or the loss of the Heide, which by now had additional value for its complement of rare species, especially of butterflies, ants, reptiles, and the occasional floral plant. In the 1990s the reserve received official permission to do the burning that they had been obliged to do of necessity. Across the country the landscape of heath had by now imploded to little more than 40,000 ha, most of it in poor health.

So they burn. They burn in late winter or early spring, typically in latter half of February after the snow has gone and before the birds breed. They like some frost. They burn in small patches, maybe 50 ha altogether a year, using fire the way a gardener might prune back a hedge or weed a flower bank. They burn shallowly, taking away only the lightest grasses and tiniest twigs, but often with enough flash heat to top kill birch and other woody invaders. These resprout or sucker from the base but then the sheep eat them back, and with their reserves exhausted, the young trees die. At the same time, to protect fire-sensitive junipers, they isolate and shield individual trees.

The Heide is as intricate as a wound-coil Swiss watch; too hot a fire might invite invasives, too broad a burn might overwhelm pockets of subspecies, too heavy a combination of fire and hoof might shatter the mosaic as fully as doing nothing might allow it to be overrun by woods. Even so, officials have yet to encompass the whole reserve with flame. In reality, weather offers perhaps 20 days a year suitable for burning, only 1 or 2 of which coincide with other calendars. In practice, that is, they burn every two or three years, and use the pattern of burned patches to circulate cattle and sheep around the reserve. Plans call for reintroducing further elements of traditional agriculture—some fields of ancient wheat, for example.

For fire ecologists from North America, in particular, the whole spectacle can seem bizarre—the scale of operations ludicrously minute, the purpose of fire restoration perverse, the operation no more pertinent to understanding the fundamentals of fire ecology than captive breeding is to theories of optimal foraging strategy for wild omnivores. This is a preserve run more on agronomic principles than natural regulation. The Dutch have seemingly drained fire ecology from the intellectual landscape as fully as they have standing water from their geographic landscape.

———————————

Yet any theory of fire ecology that aspires to something like universal status must incorporate such scenes as fully as lightning-fired wilderness in the Canadian taiga or foehn-driven flames through California chaparral.

Here is the imperial rebound: the misapplication of fire lore acquired in one setting and falsely declared normative and then applied to a very different setting.

The Strabrechtse Heide is as profound a challenge to contemporary understanding of fire ecology as the longleaf pine or sequoia was to the received wisdom of professional forestry in the 1960s. It challenges the concept of a fire regime as fundamentally derived from purely natural inputs. It challenges assumptions that fire ecology must follow from fire behavior. It forces fire from status as a mechanistic driver and into the role of a catalyst, whose effects are meaningless without reference to herbivores and other biological agents. It questions the pertinence of natural fire, autonomous from human hand and mind, as the normative point of calibration. In place of the wild landscape, to which people are added, it proposes as an alternative point of departure the anthropogenic landscape, for which wildlands are a construct created by removing people.

Of course scale matters: free-burning conflagrations in Yellowstone in 1988 burned probably four orders of magnitude more land than managers burn annually in the Heide. That's the difference in length between a foot and a mile, and in weight between a ham shank and an elephant. But the difference in proportion demands a change in type, not simply of amount; the scaling is not linear. The dappled flash fires of the Strabrechtse Heide simply haven't the space to show the fire behavior witnessed at Yellowstone, or to reveal the coarse mixture of fire effects. Both are shaped by what people do before and after ignition. Trying to explicate fire ecology at the Heide by applying the principles derived from American wildlands or the deep Australian bush is like applying Newtonian mechanics to the subatomic realm. The existing literature on fire ecology relates to an everyday world of landscapes. It is not readily scaled up to the globe or down to the garden. It is not clear that the relationships normally studied (and considered normative) can be adjusted simply to the very large and the very small. What may be needed at Strabrechtse Heide is the fire-ecological equivalent of quantum mechanics.

When foresters from temperate Europe witnessed the flame-flushed tropics and arid outbacks of their colonies, they were astonished, fearful, outraged, and finally determined to force those bizarre and blazing scenes into proper form. The result was a massive misreading of fire's place on the land. Yet North American fire ecologists are likely to make equivalent

errors amid the cultivated landscapes of Europe. They might hunt relentlessly for evidence of natural fire, even if it means pursuing charcoal laid
down at the height of the last glaciation. They will search out expressions
of fire adaptations, especially fire-stimulated germination. They will then
try to track the evolution of the current scene from such core data, and
propose that future management should seek to accentuate those pure
elements and shed 4,000 or more years of biotic contamination at the
hands of humanity and its servant species. Such a vision says very little
about the scene they are viewing, and much about the scene in which
they grew up.

In September 2008, the International Association of Wildland Fire spon-
sored a conference to celebrate the 20th anniversary of the Yellowstone
fires. The proposed topics were the ones familiar to the North American
fire community—that fire is natural; that fire's behavior determines fire's
effects; that society should accommodate itself to the inextinguishable
presence of free-burning fire since it is both unavoidable and necessary.
The idea that such a body might gather instead at Strabrechtse Heide
would seem a mockery.

Yet the inability to read the fire history and ecology of such places is a
serious failure of imagination. Greater Yellowstone and the Strabrechtse
Heide are scattered points in a constellation of the Earth's burned landscapes, and not the most extreme; any theory of global fire must embrace
them both, even if it means taking seriously the catalytic ecology of fire
in the Heide and the cultural context of the nominally wilderness conflagrations at Yellowstone. Otherwise we risk living in an intellectual world
even more contrived than those Dutch polders lying behind the grand
dikes that hold back the Atlantic.

IN THE LINE OF FIRE

Fire is fire, as the mantra goes, but fire in Korea looks very different from fire in the United States, and it reminds us how different cultures, and land use history, can result in very different strategies of fire management.

BETWEEN APRIL 5 AND 7, 2005, some 20 fires burned 250 ha and 246 buildings, forced over 2,000 people to evacuate, and trashed cultural heritage sites. The firefight lasted three days as foehn winds drove over mountains and officials mustered and coordinated resources at all scales of government: 10,000 firefighters and soldiers, 38 helitankers, 184 engines. The National Emergency Management Agency declared the affected regions special disaster zones, promising aid to assist victims, to rebuild houses, and to compensate for destroyed crops and livestock. Not an unusual event, not even a large one by international standards, but it was a fire bust that rang brazenly throughout South Korea.[1]

And its locale is what moves the outbreak from a news item to something like an apologue. The fires kindled from votive candles, military training sites, and brush burning across the demilitarized zone (DMZ). They flashed through woods that didn't exist 60 years ago. They blasted large patches through the Naksansa Temple complex, established 1,300 years ago. They burned through a landscape organized on fundamentally different principles from what most wildland fire agencies consider normative. This is a country for which wildland is a bonsai garden planted at a landscape scale. It's a country for which routine ignition comes from live-fire exercises on military bases. It's a forested country whose only fire-maintained landscape is the demilitarized zone that separates the Korean Peninsula into north and south. It's a scene that challenges typical notions of what "wildland fire" means and what options exist to manage it.

At 38,000 square miles South Korea has a land area a little larger than Indiana and a little smaller than Kentucky. Its 50 million people give it a population a little more than California and Florida combined and a little less than California and Texas. Most of the country is mountainous. Much of it began to change when Korea reluctantly signed the Khangwa Treaty in 1876 that commenced its trek into modernity. Most of it was devastated or destroyed during the course of Japanese colonization (beginning formally in 1910), World War II, and the Korean War. The war left the peninsula severed into north and south roughly along the 38th parallel. Over the last 60 years South Korea has reconstructed both its society and its landed estate, and did both along similar principles.

The longue durée fire history of Korea is not known in any detail. The modern climate arrived about 6,000 years ago, mostly temperate but within the rhythms of the Asian monsoon that encouraged dry winters, strong northwesterly winds during the spring, and summer rains. Within another 2,000 years the pines began to replace the broadleaves. The woods, or at least parts, enjoyed official protection; the Chosun Dynasty (1392–1886), for example, controlled logging and fuelwood gathering. Mostly, the small landholdings argued for close cultivation, particularly of wet rice cultivation in the valleys, but also in the mountains, even when swiddened, which made Korea another of Asia's garden societies. Instead of free-ranging livestock, inviting broadcast burning, farmers had a goat or cow that they would tether for grazing. A plausible picture is one of routine, small-plot burning for shifting cultivation and stubble, and maybe patch burning for pasture, with few occasions for far-ranging fires. Yet the chronically unsettled politics of the peninsula led to coups, wars, and unrest that from time to time removed the tending hand and created opportunities for more explosive fires.[2]

All this changed with formal Japanese colonization in 1910. Japan saw Korea's old forests as industrial material, a pattern of consumption that quickened during World War II. The Korean War widened the destruction, not least through firebombing; then crash programs for economic modernization completed the degradation. By 1960 forest stock was estimated at 9.6 m³/ha, mostly Korean red pine, a hardy pioneer. Construction timber was scarce. Fuelwood shortages caused acute hardship. Hillsides eroded. Mountain villages were beggared. The reconstruction of South Korea would involve nature's economy as well as society's.[3]

It began with state-driven investments in infrastructure. Replanting began in the 1960s with President Park Chung Hee himself planting seedlings. Often temporary terraces had to be created and soil carried up hillsides. Systematic programs began in 1973 with a 10-year Forest Rehabilitation Project that aimed to establish fuelwood plantations and prevent erosion with fast growing larch, birch, and pine. The program completed its goals in six years with the reforestation of 1.08 million ha. The adage in emergency medicine is to stabilize, then transport. For emergency forestry, this translates into stabilize, then evolve.[4]

A series of successor plans followed, each adding some complexity to the scene. The Second National Forest Plan laid out 80 large-scale commercial plantation forests with a mix of species over 1.06 million ha. The Third plan moved into "multifunctional forests" with an interest in reconciling production with public amenities. It empowered the Korean Forest Service to oversee 32,000 ha of commercial forest and over 3,000,000 ha of forest land for watershed, wildlife, and recreation. To preserve its new woods the Republic of Korea planted woodlots overseas and maneuvered to import timber that it would process and then resell (often back to the source nation). The Fourth plan, which ended in 2007, transitioned to a more sustainable forest that mixed commercial products with public amenities. The mix of species expanded, the afforestation in strips and dappled patches of tree types such as birch and Mongolian oak, spinning green fuelbreaks and rudely crafting mosaics, along with a greater reliance on natural reseeding. Meanwhile economic growth helped fragment forests with croplands, ski resorts, mines and quarries, and golf courses. The Fifth plan envisioned a "green nation with sustainable welfare and growth." Production forests would balance with recreation forests, and

Korean usage will be offset by overseas plantings. Along with the plans unrolled a series of forest laws to harden the changes in South Korean society if not its land.

It was a formula for fast yet staged development, a compound of the urgent with the logical, a kind of Asian fusion of landscape as South Korea raced into modernity at breakneck speed. The practices and discipline that made South Korea a developed country in a handful of decades equally transformed its mountain forests. In 50 years South Korea's forest stocking skyrocketed more than an order of magnitude, from 9.6 m³/ha to 125.6. Forests now cover 65 percent of the national estate.

With the return of forests came the prospect for the return of forest fire, particularly with the transfer from silvicultural plantations to multi-purpose and amenities landscapes. On April 23, 1996, serious fires roared back. The largest since the war broke out in Goseong when the military disposed of TNT on a firing range and the resulting flames ran over 3,762 ha, 227 buildings, and 55,423 "agricultural machines." Others, less savage, flared across the East Coast. These are not large fires by Big Four standards, but they are big on the scale of South Korea, and they burn with heavy symbolism. The scorched lands were restored by the familiar techniques developed over the previous 25 years. The outbreak also prompted a national discussion about what threats the maturing fire scene might hold.[5]

The primary emphases were to restore and protect. South Korea invested heavily in firefighting technologies, not only in pumps and helicopters but in research projects and fire danger rating software. There was little sense that restoring fire might also be a part of restoring fire-adapted forests. South Korea had its own logic of needs, and it turned instinctively to security forces for rapid detection and attack. Then in 2000 the East Coast fires rambled over 24,000 ha and forced the Uljin nuclear plant to shut down. Restoration followed, though the strategy favored more natural regeneration and this time appealed more to indigenous species rather than exotics. The Korean Forest Service relied on fire suppression apparatus, along with fire prevention programs, to hold fire to acceptable limits. Research emphasized fire control.[6]

What was clear, however, was that the further maturation of the Korean mountain landscape would trigger yet more fires and might, at some stage, even point to a more nuanced agenda of fire management.

The Korean War climaxed a half century of trauma. The demilitarized zone that froze the line of conflict has inscribed a chasm through time as well as space. The contrasts have become more extreme with each decade. Today the differences are visible from satellites. Look at evening lights and the north is a dark patch amid the bustling lights of northeast Asia. Look at daily hot spots and the north holds nearly all of them, fenced within its eremitic state. South Korea made what fire historians are beginning to call the pyric transition, the shift from a reliance on biomass fuels and landscape fire to the burning of fossil fuels and lithic landscapes. North Korea did not. The south has the abundance of combustion. The north has fires.

The transition is a natural trend that accompanies industrialization. But it has peculiarities according to place, time, and culture. It depends on sources of fossil fuel, a capacity to distribute its power widely through society, and the removal, forced or voluntary, of those peoples who are living in traditional ways in the countryside. In the usual scenario the transition begins with something like a fire orgy as new fuels and ignitions mingle promiscuously with traditional landscapes and kindlings before technological substitution and outright suppression lead to a dearth of open flame, an ecological fire famine.

South Korea made a deliberate decision to speed that process along. The postwar fuelwood shortage pushed it to find alternative sources of energy. Interest in quickening industrialization led the state to encourage the depopulation of the mountains by removing people to New Villages or large cities. When people left, so did the traditional sources of ignition and the purposes of burning. Agricultural fire of all kinds was legally banned in 1980. Fuels built up; fires did not. The population explosion of abusive burning that typically characterizes the pyric transition came during the war. What remains is a landscape whose population of fires now falls below ecological replacement values.

Some traditional sources of ignition persist in the guise of candles lit in temples and lanterns on gravesites and the occasional debris burn on the outskirts of towns, and when the spring winds blow, those pilot flames can rise up and blast over the countryside. But most ignitions emanate from modern conditions. In South Korea this means the military. Live-fire training leads to fires. Ordnance disposal leads to fires. Even the use of incinerators on bases has led to fires.

In the pyric transition the most dangerous time is that phase when old and new mingle without regard to environmental logic. Yet this is exactly the geopolitical and ecological circumstance frozen by the DMZ. To maintain an open field of fire North Koreans routinely set burns when the spring winds howl from the northwest and then let those flames rush south. When they strike the southern border, they kindle firefights as South Koreans try to contain them before they spill over the border and damage. The upshot is that, paradoxically, the DMZ features the only fire-sustained biota on the peninsula and is probably the closest approximation to the pre-20th-century landscape.

To Western eyes the Korean fire scene can appear otherworldly, as though transported to a planet organized on different principles. There is little pertinence to fire-dependent biotas when the nominal wildlands are planted; when wilderness is a socially meaningless term; when ecological integrity refers to an ecosystem that is a built by human labor devoted to creating terraces, hauling soil, and planting mature trees; when traditional burning refers to such relic practices as lighting lanterns in small graveyards; when there are almost no natural ignitions; when the closest approximation to the wildland-urban interface is a Buddhist temple embedded in the hills. Deliberately setting fires, even if prescribed, can seem suspicious in a security state that is still technically at war. Natural fires, managed wildfires—these are existentially blank concepts. A fire-renewed ecosystem means one replanted after burning.

The only reasonable response for the foreseeable future is to suppress fire, and to do so with massive, quasi-military force. At the VI International Wildland Fire Conference held in Pyeongchang, the Korean Forest Service staged a demonstration of its firepower by flying phalanxes of heavy helicopters to douse a simulated blaze. In time, as a more syncretic biota emerges, if tensions across the DMZ dissolve, if that other imposed divide between Korean nature and culture—between storks and Samsung—fades, there may be a place for patches of traditional burning, but it will come with a modern version of cultivation, of landscaping for purposes and according to aesthetics probably alien to the notions of the Big Four nations whose fire establishments have evolved to handle free-burning fires on vast bushlands and big backcountries.

For now, the North Koreans burn. And when they periodically declare their bellicosity by threatening to subject Seoul to a "sea of fire," that metaphor can have an unsettlingly literal referent.

———————————

For now, too, the emblem of South Korea's fire scene is the Naksansa Temple overseen by the Jogye Order of Korean Buddhism. Part of the postwar reconstruction of the Korean landscape involved cultural sites, of which Buddhist temples constitute probably a third. Nearly all lie in the mountains. The villages are gone or modernized, no longer a routine source of ignition. The wood-construction temples and the ancestral graveyards remain, still reliant on candles and lanterns, and so occasionally prone to fire. The temples also suffer from landscape fires that crowd into their surrounding woods. Today, nearly 53 such fires occur each year. Here is the Korean equivalent to America's I-zone, and like the DMZ, the uneasy border cannot be relocated or erased. The friction between temple ground and surrounding woods is fundamental to the setting. The border will persist.

The Korean solution shows the kind of synthesis that has become a hallmark of, say, K-pop. Koreans rebuild damaged temples with more modern, less fire-prone materials. They restore the woods with less fire-prone species, and where pines remain a cultural preference, they plant mature trees and meticulously clean up the surface fuels. They devise technological solutions to the problem of artifacts like the Naksansa bronze bell, such as elevators triggered by heat and smoke that will automatically lower the treasure below ground when fire threatens. They rely on rapid, massive response to quench flames.

New and old, an Asian architectural fusion. The substances differ. The form endures. The operative aesthetic is not untrammeled naturalness. Famously, the Buddha himself had a fire sermon in which he appealed to landscape fire as the very emblem of a chaotic world driven by fiery passions that had to be quelled to achieve nirvana. That is not a bad approximation of what the Land of Morning Calm aspires to not only for its temples but for its future. For the coming decades South Koreans will actively quell such outbreaks by whatever technological power they have. So long as they remain in the line of fire, that formula is unlikely to change.

PORTAL TO THE PYROCENE

And where two raging fires meet together,
They do consume the thing that feeds their fury.
—WILLIAM SHAKESPEARE, *THE TAMING*
OF THE SHREW

A commentary on the startling fire that burned in Fort McMurray, and its potential use as a metaphor of our Two-Fires era.

THE IMAGES ARE GRIPPING. Horizons glow with satanic reds squishing through black and bluish clouds, as though the sky itself were bruised and bleeding. Foregrounds bristle with scorched neighborhoods still drifting with smoke and streams of frightened refugees, a scene more commonly associated with war zones.

But we've seen this before. Big fires are big fires, and one pyrocumulus can look pretty much like another. Communities with homes burned to concrete slabs, molten hulks of what once were cars alongside roads, surrounding forests mottled with black and green—these are becoming commonplaces.

What strikes me most about the Fort McMurray images is the mash-up of foreground and background, the collision of free-burning flames with a fossil fuel–powered society. The first form of burning dates back to the early Devonian, when life first colonized the continents. The second tracks the Anthropocene, when humanity changed its combustion habits and wrenched the Earth into a new order. At places like Fort McMurray the deep past and the recent present of fire on Earth rush together with almost Shakespearean urgency.

The plot is old, the stage setting and cast of players updated.

Monster fires are no stranger in the boreal forest. It's a fire-ravenous biota that burns in stand-replacing patches. This is not a landscape where misguided fire suppression has upset the rhythms of surface burning and catapulted flames into the canopy. They've always been in the canopy, and everything has adapted accordingly. White and black spruce and jack pine and aspen experience exactly the kind of fire they require.

How big those patches get depends on how dry the fuel is, how brisk the winds, and how extensive the forest. In northern Alberta there is not much to break a full-throated wildfire. The Chinchaga fire started on June 1, 1950, and burned across northeastern British Columbia and most of Alberta until October 31, a total of 3 million acres. That single burn was embedded in a larger regional complex that probably summed 4.8 million acres. Comparable fires date back as far as we care to look.

Nor is a burning city a novelty. In North America the wave of settlement in the 18th and 19th centuries paralleled a wave of fire. The surrounding lands were disturbed, and frequently alight with both controlled and uncontrolled fires. The towns were built of wood—basically, reconstituted forests. The same conditions that propelled fires through the landscape pushed them through towns.

Only a century ago did those urban conflagrations finally quell as urbanites turned to less combustible materials; fire codes and zoning regulations organized buildings in ways that discouraged spreading flames; fire services acquired the mechanical muscle to halt blazes early; and the wave of settlement flattened. Over the past century modern cityscapes have needed earthquakes or wars to overcome those reforms and sustain conflagrations.

Then, they began to return as a broadly rural scene morphed and polarized into an urban frontier of wildlands and cities that faced one another without intervening buffers. The middle, working landscapes, like the middle, working classes, shriveled at the expense of the favored extremes. In 1986 the term *wildland-urban interface* appeared. It was a clumsy, dumb phrase, but it referred to a dumb problem. Watching houses, and then communities, burn was like watching polio or plague return. This was a problem we had solved, then forgot to—or chose not to—continue the vaccinations and hygiene that had halted their terrors.

Initially, the problem appeared as a California pathology. But it soon broke out of quarantine and has spread across western North America.

The prevailing narrative held that the problem was stupid Westerners building houses where there were fires. Most of the vulnerable communities, however, are in the Southeastern United States, and if climate change modelers are correct and such outbreaks as the conflagration at Gatlinburg, Tennessee, prove prophetic, we will see the fires moving to where the houses are. That will make it a national narrative. In truth, the problem is international, each country with its own quirky combination of fire-quickening factors. Mediterranean France, Portugal, Greece, South Africa, and Australia are experiencing similar outbreaks. North America has no monopoly.

It's tempting to appeal to climate change as the common cause. Yet the burning bush and scorched town are joined not just by global climate change but also by a global economy, and, behind both, by a global commitment to fossil-fuel firepower. That makes the issue both more pervasive and, paradoxically, more amenable to treatment. It means that, while there is one grand prime mover, there are many levers and gears. Fire is a reaction that takes its character from its context. It's a driverless car barreling down the road, synthesizing everything around it.

The enduring images of the Fort Mac fire may, in fact, be its cars. Car-propelled flight, cars stranded for lack of gas, cars melted in garages, evacuation convoys halted due to 60-meter flames, relief convoys laden with gasoline. It isn't only what comes out of the tailpipe that matters, but how those vehicles have organized human life in the boreal. The engagement (or not) with the surrounding bush. The kind of land use that cars encourage. The kind of industry that must develop to support those cars. The kind of city that such an industry needs to sustain it. The oil and tar sand industrial complex that has shaped the contours of modern Fort McMurray is in turn shaped by the internal-combusting society it feeds.

So there are really two fires burning around and into Fort McMurray. One burns living landscapes. The other burns lithic landscapes, which is to say, biomass buried and turned to stone in the geologic past. The two fires compete: one or the other triumphs. At any place the transition may take years, even decades, but where the industrial world persists its closed combustion will substitute for or suppress the open flames of ecosystems. The wholesale transition from the realm of living fire to that of lithic fire may stand as a working definition of the Anthropocene. Once parted they rarely meet.

At Fort McMurray they have collided with unblinking brutality. Wild fire burned away controlled fire. The old fires have forced the power plants behind the new ones to shut down and their labor force to flee. It's like watching an open pit mine consume the town that excavates it. It's tempting to regard the incident as a one-off, a freak of a remote landscape and a historical moment. But those collisions are becoming more frequent.

That's not the deep worry, however. The deep horror is that the two fires may be moving from competition into collusion. They are creating positive feedback of a sort that makes more fire. Those images of fire on fire are the raw footage of a planetary phase change, what might end up as a geologic era we could call the Pyrocene. They will continue until, as Shakespeare put it, they "consume the thing that feeds their fury."

Disaster is not always tragedy, and Fort McMurray and the industrial complex behind it may well escape lethal consequences. So if Shakespeare seems too elevated, consider Edna St. Vincent Millay.

My candle burns at both ends;
It will not last the night;
But ah, my foes, and oh, my friends—
It gives a lovely light!

We have in truth been burning both ends of our combustion candle, and if its light seems more lurid than lovely, there are yet texts to be read in the awful splendor of its illumination.

BURNING BANFF

When I began my research into Canadian fire history, Parks Canada volunteered a pack trip to Banff National Park as its contribution. Our trek evolved into a distilled fire history.

THERE ARE FIVE OF US, plus three pack horses, and we are strung along a trail that threads into Banff National Park. Banff is to the Rocky Mountains what the Grand Canyon is to the Colorado Plateau. A pack trip through its knotted peaks is the equivalent of a float trip down the Colorado River. We enter the park along the Red Deer River in the northeast.

Its critics often dismissed Banff as a trash park—savaged by transcontinental highways and a railroad, the Bow Valley in particular deflowered by spas, golf courses, ski resorts, swarms of tourists, a hydropower dam, its landscape degraded beyond redemption. In the mid-1990s Banff was even threatened with delisting as a World Heritage Site. Its defenders, however, note that the park has preserved nearly all its biotic pieces and holds intact its majestic matrix of streams, forests, storms, and slashing peaks. It yet retains its grizzlies, wolves, mountain lions; its elk, moose, bighorn sheep, mountain goats; a monumental megafauna to match its monumental scenery. Most spectacularly, nearly alone among Canadian parks, and rarely for North America, Banff has nurtured a habitat for free-burning fire.

A pack trip is thus a traverse through some of the most interesting fire management in North America. Banff is Canada's first national park; a century later it had become for Parks Canada the flagship for an aggressive policy of ecological integrity in which free-burning fire was the vital spark. Ecological integrity aims to keep all the parts and processes of a biota and to grant them a suitable structure so that they can maintain themselves indefinitely. It contrasts with other preservationist philosophies by ignoring such standards as naturalness, wilderness, or historical authenticity, which may or may not contribute to the perpetuation of species and how they live. A policy such as Banff's is, as postmodernists like to mutter, a contested matter.

All the themes are here: Banff is where they arose and where the relevant ideas took to the field to decide the issue. That makes Banff typical, or prototypical. What makes it special is that ecological integrity can apply to any landscape; at Banff it applies to an extraordinary menagerie of big animals and the habitats that sustain them. Fire matters because fire seems to be essential to those habitats. Ecological integrity may only succeed if Banff burns. The trick is to see that it burns properly. And that is the purpose for this curious expedition, an intellectual inspection.

A pack trip seems a particularly suitable venue because the infrastructure for packing, along with its craft, parallels that for fire. The same backcountry requirements for cabins, trails, and routines can serve both purposes. A horse broken to saddle and pack is not unlike a controlled fire. A program of burning will demand the kind of commitment, and tradition, and practical skills that keep horses on the trail. Wardens likely to expend an annual holiday venturing through the backcountry, throwing diamond hitches, hefting panniers, and chasing escaped mares through a dewy meadow, may be precisely those personalities most likely to pursue flame with tenacity, verve, self-mocking humor, care for craft, and granite-faced boldness. This particular party includes the two men most responsible for overseeing Banff's fire program along with Parks Canada's chief scientist. Whatever else they see, they will find evidence of fire or its glaring absence on all sides, and whatever else they may discuss, they will always talk fire.

The trail passes a sign announcing the park boundary. Beyond it, for miles along the valley of the Red Deer, the forest has burned.

It has taken us a day to reach that border. The distance to the boundary is not great, but the effort is—a long day of preparations; waiting, scouting, sorting, gathering and testing tackle, picking horses. We rode for only a few hours up the narrowing Red Deer to Barton's Camp, a private outfitter's resort, near the border.

A pack trip is not a casual enterprise, and Banff's tradition of backcountry horse patrols demands an elaborate infrastructure of trails, stock, lore, and cabins. Such patrols were the means by which early-day wardens observed for themselves what was happening in the park they oversaw. They watched for poachers, beetle outbreaks, and fires. For a while, the cabins evolved into permanent residences before reverting back to temporary way stations. Throughout, the practice of patrolling endured. On a horse you smelled as well as saw. You felt the land in ways that viewing it through windows of aircraft or autos did not allow. The patrol was how, in a vast and looming landscape, you made your presence known.

The program persists, if less visibly, amid a modern world of helicopters, snowmobiles, personal computers, pickups, and SUVs. Appropriately our journey thus begins somewhat outside the park proper, at Ya Ha Tinda Ranch, probably the only east-slope valley along the Rocky Mountains that has not succumbed to suburban and second-home sprawl. The ranch itself sprouted on the old Brewster spread when in 1918 three park wardens built a cabin to support their winter patrols. The ranch proved ideal for raising the horses the park required. Over the next two years the local wardens relocated the house and established a horse-ranching operation that today boasts 200 stock. Yet that facility is only a down payment. The horses require trails; the trails lead to cabins and corrals; the stock demand feed, whether as imported oats or native pasture; and the warden service must retain a staff for all of this. Packing demands unique equipment, specialty skills, and training—Banff puts its warden recruits through a two-week course in packing and patrolling.

At last we have wrestled horses, panniers, saddles, and riders into some semblance of a pack string fit to ride through the valley of the Red Deer. The trail wends through young pine, along a mix of abandoned roads, animal paths, and surveyed lines for seismic sensors.

———

Cliff White leads. He does so not by any rank but by the sheer bustle of his enthusiasm, and that is how he has led the Banff fire program. He had grown up in Banff—knew its marvels and mishaps firsthand, relished its outdoor sports and its wildlife, knew its personalities and politics. After high school, he spent a year at the University of Calgary, then dropped out and spent much of his time skiing, particularly in the United States. In December 1974, while in Missoula, he heard a lecture by Bob Mutch, then a researcher with the Intermountain Fire Sciences Lab. The Forest Service was slowly rousing itself into a new era of fire—had, as of two years earlier, experimented with allowing natural fires to burn freely in the White Cap Wilderness. Mutch's ardor proved infectious. Cliff enrolled at the University of Montana and cobbled together a major that combined wildlife with fire. He then headed to Colorado State University, at that time one of the few schools that offered graduate courses in fire, and worked through a master's, a fire history of Banff.

He had his mission; and his timing was impeccable. In 1979, as he returned to Banff with an appointment as a warden, Parks Canada was restating fire policies. While Banff's aboriginal inhabitants had used fire, its park protectors sought to abolish burning as environmental wreckage and human intrusion, and to a remarkable degree they had succeeded. They had done so with such thoroughness that through disuse Banff's once-vaunted fire suppression organization had withered into insignificance, staffed, if at all, by temporaries. Fire management seemed an archaic practice, like branding and blacksmithing a craft of largely antiquarian interest.

But thinking changed. Preservationist philosophy argued for encouraging natural processes like lightning-kindled fire, and ecologists began to tally the ways fire's removal had inflicted its own damages on the land. The argument was not simply that fire suppression had allowed fuels to pile up into conflagration-stoking horrors, that controlled burning was needed to thin out that wild woodpile to make fire protection easier. Rather, the core thesis was that fire belonged biologically, as much as bighorns and marmots. Trying to abolish fire was akin to abolishing snowstorms. It simply made no ecological sense.

So Parks Canada resolved to allow fire to return. That meant Banff needed a dedicated fire program, one that could fight fires the park didn't want and light those it did. What this meant in practice was unclear. Cliff White told them. "Cliffie" sparked Banff's moribund program back to life. By 1983 Banff boasted a first-rate initial attack crew. A year later it was staging its first prescribed burn. Two years later, the park's centennial, Cliff published a historical survey of fire in Banff, a model for assessing fire's place. Parks Canada acknowledged Banff as a national prototype. That founding fire crew became the nucleus for the crack fire command teams that the agency dispatched to major wildfires—bad burns it wanted extinguished that required talent beyond what an individual park has in-house. More permanently, the Banff brotherhood dispersed throughout the system. Wherever they landed, they kindled modern fire programs.

When he began, Cliff had little to go on, and endless reasons not to bull ahead over the bureaucratic muskeg. The earliest trials, at Two Jack Campground, stirred a fire whirl of controversy. The public didn't like the cutting that had to precede the burns, and they most certainly did not approve of the fires. Had they understood the full outcome, they might have approved even less. The plans called for a severe fall burn. What happened was, the resident lodgepole failed to seed properly, those that survived the fire were hit with pine beetle, the elk gobbled up the aspen suckers, and the place, against all expectations, came back to spruce. Such circumstances, Cliff learned, called for a fast, hot spring burn. The program moved on.

Year by year, Banff underwent its institutional chrysalis. Cliffie did it with sheer personality; with unbounded zest, a high tolerance for controversy, a remarkable capacity to absorb information and ply it in the field. His impact was less charismatic than catalytic. He is the behavioral incarnation of mixed metaphors. His wiry frame is as restless as a gerbil; ideas fly from him like sparks. He speaks rapidly, in a clipped vernacular. He "talks" in the same way a hummingbird "flies." He bustles from one project to the next, from one idea to another, the way a hungry omnivore might prowl through a landscape. Fire management, like life, was an experiment that never replicated itself. When he learned that a large burn was succeeding, he famously replied, Well, that worked. Let's try something different.

———

This is contested land—has been for millennia. Aboriginal Americans fought over the bison that spilled into the Rockies from the plains. Contemporary Canadians quarrel over fires. A gravel road that connects Ya Ha Tinda to Banff proper passes two burns, one on each side of the park boundary.

The Dogrib Creek fire likely sprang from a hunter's campfire on lands protected by the Alberta Forest Service (AFS). It was mid-October, the site was remote and mountainous, so the AFS loose-herded the burn and waited for winter to snuff it out. Instead a chinook wind spilled over the divide, hurled sparks across the valley, and propelled the flames like an avalanche into a downslope run that splashed through the foothills. The AFS mobilized for a firefight. They bulldozed one line across the fire's head and watched the flames hurtle over. They bulldozed another, and watched showers of embers skip across. And so it went until, finally, the winds died, the fire calmed, and the AFS staffed up for massive salvage logging, as though intent to punish the forest for daring to defy it.

Over the mountain, starting roughly at the same time, Banff had a fire of its own. This one they had set, nestled in the Bare Range. They ran it into the tree line, letting the steep terrain, rocks, and wind box it into a corner. Then they watched, as across the summit, outside the boundary, a phalanx of feller-bunchers slicked off charred trees like pond scum.

———

It takes time, and patience, and talent, to ready a pack string. This is our first morning and we must, as a group, learn the drill. Yesterday we managed to load everything from mounds of goods dumped out of trucks and to match horses with riders and pack stock with tackle. Now we must be more precise.

We collect the horses from the corral and slip on halters and lead them to the tack shed where the blankets, saddles, bridles, tarps, ropes, and pack boxes (panniers) are stored. The riding horses get saddled first. Then we place the packsaddles on Rocky, Banjo, and Ziggy. We work in pairs, one man to each side, each hoisting up a matched tack box. The packs must balance. Stray gear in duffle bags goes on top. A tarp covers the lot, and an ineffable diamond hitch seals the tented mound tight. Then we slip bridles onto our horses and direct them to the trail. Cliff White leads, with Steve Woodley and Ian Pengelly behind, each with a halter to a pack animal. Cliff's father, also named Cliff, and I follow.

It is an easy morning's work. The camp is a private facility, and it supplies our cabins, meals, a corral and feed for the stock. It's a morning for the novices to begin their apprenticeship and for the veterans to rekindle their skills. The most experienced among us check cinches and eye the heaped packs and help soothe the nervous stock. Halfway through loading Rocky, they decide the packs aren't aligned right, and undo the pack boxes and retie them.

We cross the park boundary and enter a burned forest. The valley narrows here, before widening again. We ride on the north side of river, trekking across the south-facing slopes, those most easily burned, and which were in fact fired in the autumn of 1994. Planning for that burn had begun seriously in 1991, barely a decade after the park's fire program had been rechartered. It required that Banff's fire crew do something unusual and difficult: it required they refit skills developed for suppressing fires into starting them. Typically, emergency fire crews respond poorly to long-return projects; the turnover of seasonal crews is high, crewmembers suffer from withdrawal of adrenaline rushes, they grumble over routine labor. Not least, their image of themselves must change from crisis fire fighters to calculating fire lighters.

Along the Red Deer Valley preparations included collecting weather data, estimating fire behavior from prevailing winds and terrain, and the cutting of a fireguard—a fuelbreak of thinned woods 100–200 meters wide—along the border itself. Cutting, piling, burning: the fireguard took a year to complete. Its purpose was to contain the kindled fire within the desired borders, acting as a kind of loose fuel fence. For the actual burn, a system of hoses and pumps drafting from the river was erected to help quench the flames as they approached the prescribed borders. (Before the fire ended, the network of hoses extended over two miles.) A spur ridge on the south side of the river was also ignited, as a kind of fire jetty against future spread. The burn worked more or less as planned and blackened over 1,500 hectares.

It was a bold act. If the fire had bolted beyond the border, it would have devoured outfitting camps, threatened the Ya Ha Tinda, and perhaps provoked Alberta's fire authorities into a ruthless retaliation. Yet it was deemed not merely useful but necessary. Parks Canada policy stipulates that its units burn at least 50 percent of the acreage known historically to have burned. What appeared extensive was roughly Banff's required annual tithe. But there were other sites less tricky than the steep valley of the Red Deer where one could have satisfied bureaucratic quotas. The Red Deer burn had other purposes.

The park, specifically Cliff, would like to reintroduce its vanished bison. The reasoning is both simple and subtle. The bison were once here—the evidence is everywhere. Almost certainly the beasts filtered into, or were driven deeper into, the interior of Banff, and the valley of the Red Deer is an obvious corridor. Heavy snows would then seal them from escape back to the plains, like a fish trap. With the bison caught inside their mountain corrals, natives could hunt them throughout the winter. Relict bison bones litter the old route like those of oxen along the Oregon Trail.

A bonus to a bison revival, so Cliff reasons, is that they might help "float" a wolf population still rebounding from historic lows. But there is little prospect to herd bison through a Red Deer Valley that has become a tangle of conifers, as thick and prickly as rolled barbed wire. It is hard today to imagine the south-facing benchlands as an ancient wildlife corridor for gaggles of shaggy megafauna. To restore such conditions one would have to thin the revanchist woods to the point that they became prairie peninsulas once again. The most direct strategy would be to burn them back to their aboriginal past. Cliff is convinced this is doubly right, because those fresh-burned lands, ripe with prairie-grass proteins, were, in truth, bait exploited by bison hunters to lure their prey inward.

Eight years later the landscape still bears witness to the burn. Black and silver snags stand like spikes; the forest floor is a coarse weave of fallen logs and rank grasses, forbs, and shrubs, impervious to erosion. Ian, Steve, and Cliff agree the place is primed for another burn, probably several, because a fire program is not a one-off event. Like backcountry patrols, it's a lore, a commitment, a relationship. Banff's burners must return, just as its packers endlessly revisit the great web of Banff trails.

━━━━━━

For a couple of days we are joined by two backpackers, one of whom is Mark Heathcott, formerly of the Banff fire program, now on a working holiday from his Calgary post as western fire coordinator for Parks Canada. We round a bend and ford the river and find them electric with excitement, a small fire burning on the trail. Not 15 minutes earlier, Mark blurts out, they saw an elk dash across the trail, followed by a calf, followed by a grizzly in full gallop. The calf swerved, the bear matched it in stride, and they watched it overtake the calf, and with its paw on the fallen head stare at them to see if they wished to contest him for the kill. They didn't.

Banff has what few fire-rival reserves do: a multistranded conceptual rope that knots combustion with creatures. Big fires and big animals, that is what justifies the risks and expense of its fire program. Its animals need fire, but it is equally true that its fires need those animals. An abstract argument that fire belongs for reasons of ecological purity has almost no constituency. But if fire is linked to the survival of charismatic creatures, it can rally supporters. They may not like fire as fire, but they will accept it as a necessary shaper of habitat for the animals they wish to preserve.

Theories of nature's economy and how to manage always tack closely to the wind of cultural perceptions. Banff follows what Cliff White calls a predator model of ecology. In this perspective, the big animals do not simply scrape off a surplus crust of biomass: they shape the entire structure of the biota. It's a top-down paradigm, not particularly in academic favor over the past few decades in which bottom-up models have struck a more socially responsive chord. It seemed, instead, that the little folk, the unnamed masses, the silent toilers in field and factory paradoxically determined the fate of nations, and that the proper study of man was not man—certainly not great men—but all the throngs of ethnicities, races, working classes, and genders of humanity's massed spectrum. History's Gaia depended on its unseen human bacteria, algae, and microbes, the "creatures without history."

But the academy's model of the world became increasingly out of sync with political reality. In an era of predatory plutocracy, when a single rogue trader can topple a world bank, when a pack of currency speculators red in tooth and claw can unhinge exchange rates, when carnivorous CEOs can destroy a trillion dollars of stock market value, plunge huge corporations into bankruptcy, and throw the toiling minions into unemployment lines, a predator model seems more plausible. Cliffie believes Banff has ample evidence within its own microcosm. Its big creatures have shaped the world for its small. It needs its megafauna, and those megafauna need fire, the ecological equivalent of Schumpeter's concept of capitalism as creative destruction.

We cross a bridge that spans a tributary to the Red Deer, a stream deeply entrenched into hard rock. Beyond, the valley opens. Bog birch ripen into an immense meadow, itself surrounded by a wooden fence. We cross through the gate and turn north to Scotch Camp, nestled along the forest fringe.

We unload our stock—the pack animals first, next to the cabin so we don't have to heft the boxes far, and then lead the horses to the corral where we remove their saddles for storage in the tack shed. Our riding horses follow; saddles, bridles, halters. The small corral opens into the grand, fenced meadow. Ian places bells around the necks of Ribbon, Rocky, and Gringo, enough to track the herd, and opens the gate. The horses dash to a small stream for a drink, then prance into the meadow in a clanking cavalcade to feast on the ripe grass. A yearning for oats will draw them back, as a group, come morning.

To the west, Cliff notes, there is a wolf den. We should hear its residents in the evening, and if we are lucky, we might be able to observe them as they hunt. The moon looms high and bright. We hear yaps and howls through the night.

Scotch Camp, like its backcountry cognates, has its code. You leave the place as you found it. Whether you hike, ride, or lead a pack string, you depart with the cabin spanking clean. The only record of your presence should be your entry in the logbook.

On entering, we unlock the metal bars across the doors and shutters. We ignite the pilot flame on the gas stove, and toss in kindling for a wood-burning stove nearby. Mostly we use the gas stove to cook, the wood burner for heating the cabin, warming kettles of water, and disposing of trash. When we leave, we will wash and put away any dishes and utensils; stash sleeping pads and blankets in their closets; sweep and mop the floor; replenish the kindling; lock up the cabin and bar its windows. The prescription is inflexible. If everyone follows it, the cabin network is self-sustaining. The logbook and radio call-ins ensure that everyone knows who has lived up to or failed the code.

All this too takes time. Ian and Cliff wake early and entice the horses back into the small corral with buckets of oats. Cliff Sr. joins them, as they curry and brush our tiny herd while the rest of us prepare breakfast. After eating, we begin cleaning the cabin and saddling the stock. We load the tack boxes with the horses hitched outside the cabin. The mopping and shuttering conclude. We head into the grand meadows beneath the Mounts White and Tyrrell.

———————

The grassy lake is fringed with scorched trees and pockets of burned snags. No less than Scotch Camp, the meadow is part of a network, and it requires constant tending. Aboriginal Banff did it with a pragmatic mix of burning, hunting, and bison grazing. When that ended, the valley went to woody seed. For decades, its fiery weeding in particular has been ignored and trees have crowded into it from the mountain flanks.

The new fire program has attempted to revive that unwritten code. It burns to drive back the strangling trees, thin them out, stimulate a richer flora from the meadow. The last big burn occurred in 1996; another, smaller one raged last year, and that is the source of the scorch-killed needles. The grassy cover makes it relatively easy to kindle fires. But the biotic grunge that has accumulated in the ecocorners and under the woody furniture may be more than a fire mopping can remove. It is one thing to clean, another to rehabilitate. For now, the park seems content to reinstate regular burning in the hope that flame will at least hold the scene in check. Scotch Camp is clean because its inhabitants religiously scrub it. The meadow is cluttered because they haven't.

Scotch Camp meadow is not alone. It is part of a larger system of wildlife corridors and grazing grounds. Without regular fire, conifers would close off those corridors like cholesterol plaque pinching off an artery. Probably the park has not burned enough to scour them properly—there is enough fire to hold against the woods, not with enough to drive them back. The narrower valleys like the Red Deer have absorbed harder fire hits, yet still cry out for more.

———

Approaching Scotch the day before, we had paused beside Coyote Creek to examine an exclosure, a high metal fence designed to exclude elk. The area had burned in 1994; the exclosure went up two years later. The contrast is absolute. Within, the pen practically chokes on aspen, a meter or two high. Outside, there is nothing, save what, on close inspection, one can find of aspen clones cropped within millimeters of the ground. The burn stimulated aspen growth; the elk ate it back. If Banff wants aspen, and it does, if only for reasons of ecological integrity, it must control its elk. For that it needs a controlling agent. It needs predators.

The predator of choice is the wolf. While the Banff wolves never brushed against extinction, they did plummet in numbers, and they continue to suffer losses every winter as they follow prey outside the park. A few skilled wolfers account for most of the deaths. Getting the wolf numbers up is a means to contain the elk, who have found refuge by clustering around roads and human habitations, where wolves are reluctant to tread. Tweaking the system has proved complicated, although by the numbers it has worked. Wolves are up, elk are down, and aspen sprout profusely following burns.

But the mix is not quite right. The elk still harvest aspen more quickly than can grow to maturity. Ever the optimist, Cliff believes the program has passed its biological Lagrange point, that it is entering an era of virtuous-circle ecology. Ian, who runs the fire program, remains skeptical. He worries that the park is living off the accumulated wealth of favorable habitats stockpiled as a result of large fires in the 19th century. The best habitats, that is, are those that grew up after major burns a century or two earlier. Nothing in recent decades has approached that magnitude of burning. Today's burning may be too little too late. While young fire landscapes are appearing, old ones are dying even faster. The promised revitalization of aspen has not yet arrived.

Paths diverge, our party fissions. Cliff wants to reach Windy Cabin by another, more devious route through Lower Panther Valley. With the pack string to care for, Ian prefers the shorter, more direct trail over Snow Creek. The two groups split in the meadow.

———————

In 1987, having resuscitated the Banff fire program, Cliff White was seconded to Ottawa to help defibrillate Parks Canada's Directive 2.44. That policy reflected some of the best minds in Canadian fire. From the start, policy thinking had diverged from American models. It avoided wilderness, which the Canadians regarded as too culturally ambiguous to guide practical field operations, and it shunned a doctrine of natural regulation, a laissez-faire theory that insisted that administrators needed only stand aside and let nature run its course. Instead, the Canadian Park Service tried to craft a third way, a roughly science-based program that would aim to keep the biota intact and accept human intercession as needed. It looked a bit like corporatist Canada—eager for the trappings of a welfare state yet one that kept the predators intact. The doctrine of ecological integrity received its fire mandate when Cliff compiled a survey of park needs, *Keepers of the Flame*. Unlike the 1979 directive, the revision would argue for actively setting, not simply tolerating, fire.

The words were easy: the devil was in the doing. Cliff campaigned to get the ideas written into new directives, then to get them approved, and, no less critically, to get them funded. All that happened, much of it through the able bureaucratic hands of Stephen Woodley, who succeeded him in Ottawa, when Cliff returned to Banff as chief for conservation biology. Ask Cliff about his contribution, and he urgently explains that lots of others made critical decisions; the director general, chief wardens, superintendents. But ask others, Cliffie's friends and foes alike, and they will admit that Cliff White was the indispensable man. He never wearied, he never faltered. Even tinder, well prepared, requires a spark. Cliffie showered sparks. If one trial failed, he would try another. If one succeeded, he might try another as well.

Meanwhile, as Woodley expanded Banff's vision throughout Parks Canada, in Banff itself Ian Pengelly, another of the original Banff brotherhood, had risen to the status of park fire officer. Burning—operational, experimental—continued along proposed wildlife corridors. In 1994 the park oversaw the mammoth Red Deer fire. In 1996 crews burned the valley of Scotch Camp. In 1999 they fired Panther Valley, where Windy Cabin resides.

We cross over Snow Creek summit amid July snow that turns to sleet and, dropping into Panther Valley, to rain. Near the pass itself we ride through the lodgepole pine progeny of the 1921 fire. The forests of Banff tend to be patchy, grouped into clusters that have resprouted in the ashes of major burns. Ian lets the three packhorses follow the broad trail on their own. They jostle for their rank in line but otherwise trudge along without coercion. The pack animals are fitted with nose nets that prevent them from nibbling at grasses along the way, so they reluctantly stay with the string, knowing what awaits at the corral. They follow their food, as fire does. The sun blasts through the clouds, the weather suddenly warms.

Windy Cabin shines below, white as an iceberg, near the junction of three streams. We ford two of them and begin unpacking while we wait for Cliff and his troupe to arrive. Strewn beneath the porch windows and doors of the cabin are heavy planks studded with upturned nails to discourage prowling grizzlies.

Windy Cabin is a place for regrouping, roughly halfway through our trek. There is good pasture for the horses, a snug cabin from which to explore, a landscape rich with wolves and bears and three years of extensive, deliberate burning. Here craft and place converge.

Good packing is a practiced art. It has its order, its rhythm, its technique. You can't haul a camp in sacks tied to a saddle horn, or hold a load with a granny knot. The horses must be trained to a pack, the packs must be arraigned with balance and lashed with firmness and flexibility, a packer must know when to lead the animal and when to let go. Every horse is different, every trail has its idiosyncrasies. Loads shift, horses spook, accidents happen. A pack trip is an exercise in adaptive management.

Banff uses military pack saddles mounted atop double blankets. The saddle is precisely symmetrical, with double metal hooks on each flank. One pack box goes to each flank. Cliff and Ian work together, across a common saddle, seeking the elusive balance that is the packer's ideal. They begin with a basket loop, each man to wrestle the box on his side. A rope is fixed to one hook and looped around the other with the end allowed to fall to the ground. They pull the rope crossing between the hooks down until it makes a large loop. Then they hoist their box up and hold it with one hand or shoulder while lifting the loop around the outside of the box. Still holding the box they tug the loose end of the rope until the loop tightens, snugging against runners along the box's sides. They adjust their heights so that the boxes, of equal weight, rise evenly. Then each man pulls the loose end of his rope under the box and toward him and then over and around the cross rope into a large, loopy slipknot. The slipknot seems odd: the point, after all, is to hold the pack box securely. But horses bolt, horses stumble, packs strike rocks and branches. If something goes wrong, the boxes must be able to tumble free. A firm slipknot is the paradoxical ideal. Since we have extra gear in duffle bags, these go cautiously on top. Cliff and Ian raise and lower paired bags gently since Rocky might easily start at sudden spasms of movement and unexpected weights on his back. Then they toss the loose end of their ropes over to each other, running it through each loop and returning it, where it is tied off into another slipknot.

This much is simple. Next comes a tarp. They estimate the size of the load and trim the tarp to fit. Slowly, with soft words for Rocky, they nudge the tarp over the tied bundles. Again, they balance the canvas, calling to one another whether to pull or release. They tuck in loose ends and flaps, always with slow, deliberate movements. Now comes the lash rope with its wooden hook. Ian drapes the rope over the top of the tarp, running head to tail, and lets the loose end dangle to the ground. The other end, with the hook, he tosses gently over the pack to Cliff, who hands it back under Rocky's belly so it will overlap the saddle cinch. Ian grasps the hook, loops his rope through it and tugs to tighten. Then he tosses the rope over the top to the left of the rope he has just tightened and pulls a loop from that strand under the first. He then passes a second loop through this one, drawing from the rope he first laid out. This double looping creates a crude diamond shape and leaves four strands, two to each of the boxes, one to the front and one to the rear of each. Ian and Cliff now proceed to tighten each strand in a prescribed order. Cliff tugs on the front rope, strings it under the rear of his box and then up the box's front side and back to the top, where it is again pulled. The tugging passes to Ian who does the same on his flank. The upshot is that the diamond hitch on top helps distribute the weight of the boxes equally. Ian ties off the final end with a double slipknot, taking care to tug upward so that none of the rope loosens or that the strain becomes unbalanced.

Rocky shuffles. As they check their work, each man crossing to the other side, they place their hands on Rocky's rump to let him know where they are. Rocky and Ziggy are widely acknowledged as the "boneheads" of the pack string. Someone at the cabin thoughtlessly tosses some kindling onto the porch. That sharp retort startles Rocky, who tugs on his halter. Ian had only loosely looped the halter since the hitching post was not designed for heavy pulling. The halter breaks free. Rocky jumps away from the cabin. The pack bounces and rolls; but it holds. Ian slowly approaches the nervous horse and reclaims the halter.

We reach the final streams around Windy in good order. Along the trail we watch the packs as they slide and shift. If they move too far off center, then we stop and adjust. A tug on the cinches might be enough, or a retightened diamond hitch. Occasionally there is nothing for it but to repack; not today, however. Beyond the Windy's creeks one of the hikers is waiting for us. The water is still high, and she would like to cross on a horse. We ford the streams, and I surrender Ribbon for Ian to take back for the hiker. But after a long day's ride and chaffing from snow and wind, we are slow to secure the pack string first, only removing their nose nets. Rocky, Ziggy, and Banjo wheel about to rejoin the group. Ian shrugs. We have arrived, and if they wish to splash across the streams on their own, it matters little.

On the return trip, however, there is a commotion between the streams. Something has spooked Banjo—or Rocky, free of his net, may have nipped him. Banjo bucks wildly through the bog birch, his boxes and duffles fly free. Probably only 50 yards from the cabin Ian must capture and calm the horse, then patiently reload and tie down every article of tack. He does so steadily, Ian the Imperturbable, then leads the string to Windy, where we systematically unload and turn the horses out to pasture.

———————

The Panther Valley pinches almost shut near Windy Cabin. The mountains narrow into an S-shaped pass that segregates two valleys, Lower Panther to the northeast, and Upper Panther, the smaller, to the west. Together they account for the largest acreage from prescribed burning in the park.

Lower Panther burned in 1990. It seemed an ideal setting: the south-facing slopes were extensive, the valley more or less self-contained. A fire was unlikely to escape over the rocky summits to the north or beyond the park boundary. The prescription called for a fall burn when the heavy fuels would be dry and the winds predictably from the west. The fire burned more or less as forecast. The program could shift the next burn to Upper Panther.

Predicting fire behavior is seemingly the most rigorous aspect of fire management because it is the most readily mathematized. Fuels, terrain, weather—all can be analyzed as physical phenomena, quantified, and described by algorithms. In practice, valleys are not wind tunnels; pines, bog birch, and fescue are not blocks of carbon bullion; and winds do not align around mountains like iron filings about a magnet. The particulars make fire behavior forecasts more art than science. It is a craft, learned from long apprenticeship. The 1990 Panther burn told the Banff fire program that it knew enough to push ahead. The next year, however, a fire at Norquay in the Bow Valley escaped and Banff had to import assistance from Wood Buffalo National Park. The fallout of that blunder included a major review of the Banff fire program. The panel recommended changes but affirmed the fundamental soundness of the scheme and, better, assured the program got a major infusion of money.

They burned, and they learned. They discovered which variables dominated in what seasons and for what fuels. They burned some slopes in the fall, others in the spring. They underburned when the moisture in the canopies was high; they kindled for crown fires when the surface litter and canopies were as parched as kiln-dried lumber. They burned south-facing slopes first because they had the longest windows of opportunity. They ignited some slopes at the top, some at the bottom. They kindled some fires with spot ignitions, like drips of acid on a board. They kindled others in long streamers. They observed that some scree slopes were worthless as firebreaks because the rocks covered wood and organics that carried fire, like a smoking fuse, under the stone and into open forest. They confirmed that mostly the striking topography of the Rockies drove fires upslope, that terrain could override other considerations. Over and over, they relearned the obvious: that it was easier to start a fire than to hold one. They learned that things could easily go wrong, and Ian soon devised the distinction between a zone of ignition and a zone of containment, between what they intended to burn and what, in truth, they did burn and could accept.

The 1999 Upper Panther burn was intended to replicate the 1990 fire. It started well enough; a spring burn on south-facing slopes. But a cold front approached, and the winds poured over the glacial Sawback Range like a chinook. The fire skipped through the stony pass and entered Lower Panther, its forests underlain with snow. To everyone's astonishment, the fire blasted through the needled crowns at 100m/min, as through the lodgepole were prairie grasses. An independent crown fire—one that rushes through the canopy without assistance from surface fires—is a rare event, more theoretical than witnessed, fire behavior's equivalent to gravity waves or neutrinos. Yet the unimaginable had in fact happened. Even so, the fire had held within the park's borders.

═════════════

The environs of the Panther Valley are prime grizzly habitat. We see scat on the trail, and the meadows are virtually spaded into gardens by grizzlies digging for yellow hedysarum. Come August they will scour the hillsides for buffalo berries, their prime food for bulking up to endure winter. While berries and tubers are their principal diet, they will eat whatever they can, and they are belligerently territorial. They'll take elk. They'll fight for carrion. They'll dig out and kill wolves. They'll kill and eat other grizzlies, particularly cubs, caught in traps by radio-collaring researchers. The nail-studded wooden doormats around Windy Cabin are precautions borne of long experience.

Canadian environmental groups have combined grizzlies and old-growth woods into a common slogan, the Great Bear Forest. But they could not prove that at Banff. Virtually all the food-stocks for grizzly come from light-rich landscapes. The critical buffalo berry, in particular, grows in inverse proportion to canopy closure. The older, the denser the woods, the fewer the shrubs and the sparser the bears. Virtually all the prime grizzly habitats, that is, derive from old burns at various stages of recovery. A fire-excluded park would be a park that excluded bear.

That insight bolstered enthusiasm for the fire program. Burning was not simply about deep-ecology musings over process preservation, or about quarrels with foresters over the value of old growth. A badly wrought fire program threatened Banff's big animals, and if the park's predator model was correct, any such loss would cascade throughout the biota. The public argument would be fought over what the public considered as charismatic creatures, and in the hierarchy of predators, grizzlies probably trump even wolves. Oddly, elk often graze near wolf dens, which seems suicidal except that wolves appear reluctant to kill near the den because the carrion would attract grizzlies, and the grizzlies would soon discover and attack the den.

We spend the afternoon riding to Upper Panther. We stop at several sites where we measure the year's accretion of grizzly diggings. The association with fire is obvious yet oblique. Without fire the hedysarum would senesce, and without burns the lodgepole and other conifers would entomb the forest floor in a sarcophagus of needles and woody debris. But the route between fire and bear is as tortuous as the trail Cliff navigated around the Lower Panther. The plants have their life histories, the bears have theirs, the fires another. No simple design spans them with a single algorithm. No simple prescription can determine what fires should burn when. The interactions are infinite, rife with unintended consequences, as when the escaped Panther burn killed one of the radio-collared wolves, an alpha female.

Another group has joined us in the cabin, which has become crowded. The garbage piles up, too much for the kitchen woodstove to consume. Cliff gathers up some bags and hauls them to a concrete slab that once served as an outbuilding. He collects some branches and ignites the trash. Others gather round, drawn by the open flames and quiet crackling. We add more wood to keep the fire going even after the garbage is gone.

We have, in fact, burned all along our route. Combustion cleans. At Scotch, Mark spotted a small stack of lodgepole limbs and added to them a wheelbarrow load of chips from where he had been splitting wood. Then he set it on fire. Mark was born outside Calgary, but spent his youth prospecting and fishing in the foothills. His passion for the outdoors came from his father, a petroleum geologist who had been a star athlete in college, but his fire genes he got from his "born burner" of a mother, the product of Mormon farmers and ranchers from Oyen, Alberta. While we talked about how Parks Canada manages fire, he fussed over the slash burn with a garden rake. The parks guys are burners, he explained. That's how he fights fire, that's what makes the Parks Canada fire teams different from the more suppression-dedicated crews in B.C. and Alberta and Ontario. The parks guys will find a good place to kindle a backfire, and when the time comes they will ignite. The others hesitate. They try to attack the fire head on, or to outflank it with bulldozers and water bombers. The parks teams will burn. They prefer to fight fire with fire.

Most of them, like himself, like Ian, he notes, came out of the Banff initial attack crew that Cliff inspired. When Ian arrived in Banff in 1980, after a long apprenticeship at Jasper, Pacific Rim, and Glacier Parks, the fire program was on life support. Cliffie revived it, attracting personalities as tenacious as his own. They learned their craft. Ian stayed of course, but the rest of us, Mark explains, we've scattered throughout the system. Banff couldn't hold us.

You know, he adds, smiling, that Cliffie once burned down a house. He was about nine and he and a friend were playing with fire around an abandoned cabin near Banff and they managed to lose it and the house became ash.

Yes, things happen around Cliffie. But he's the reason the fire program exists. He's why it worked.

———————————

The day starts slowly. After the horses have been gathered from the pasture into the corral, and everyone has sat down for breakfast, Ziggy manages to pry the gate open. The gate is heavy, all of logs, and it took a good effort to hoist the end onto its resting stump. No one bothered to latch the top; it seemed unnecessary. Ziggy, however, works his head under the lower log and lifts and pushes until the gate springs open. The horses have had their oats. They dash back into the pasture. It takes another hour to round them up, even with an additional bribe of oats.

⸻

Our route to Flint's Cabin is a medley of paths, beginning with a fragment of the old road to Flint's Park, then over horse trails, a fire road, another trail, and finally over the gravel road that once joined Banff townsite with Ya Ha Tinda.

That park-spanning road had been completed, engineered and graveled, in 1962, but as Parks Canada redefined its mission from tourism and scenic protection to something approaching a doctrine of ecological integrity, the road became an embarrassment. It was the kind of development the parks should prevent, not promote. In the early 1980s, it closed, largely at the suggestion of the man who would become Stephen Woodley's graduate advisor. By 1985 mitigation measures were in place, although costs were high, intervention had its own unintended consequences, and reform did not instantly happen. The road's side cuts, culverts, and survey-straight gravel thoroughfare could easily last a century.

There are many routes by which to "manage" nature. Parks Canada had decided to close one, and take another, and Woodley assumed responsibility for planning where and how that new path might go. The options were many, but all would wend around fire. A 1996 study of Banff, in fact, identified fire as the keystone process on which the sustainability of the park might depend. Typically for government agencies fire is what remains after every imaginable constituency has claimed its tithe. Parks Canada, however, made space for flame. Whatever groups might come to the table at Banff, they would find themselves seated around a common fire. It worked because Banff's charismatic creatures needed fire and its fire characters were ready to use it.

We turn into the headwaters of the Cascade River, and then cross over a ridge and dip into Flint's Park. Cliff and Ian, riding lead, spot a mountain goat glaring at them from above the trail. They wave excitedly to the rest of us, the dawdlers, and then watch the goat plunge ahead.

Around a bend, not 50 meters away, they spy a wolf, obviously trailing the goat. Further down the trail we come upon a massive-logged goat trap, now decaying. Above us, the mountain slopes, steep and slick as only glaciated valleys can be, sport specks of white. These are small herds of goats and bighorn sheep and even, to Cliff's delight and dismay, elk. These, too, belong in Banff's biotic pantheon, and they, no less than wolves and grizzlies, find their world shaped by fire.

———————————

Trails tend to become passages for fire; our route to Flint's Cabin is no exception. The evidence of the 1936 fires abounds everywhere in the texture of the forest—its even-aged blotches, its stringers, its sentinel snags. At Cuthead Creek the peculiar patchiness of the conifers tells of the 1914 and 1929 fire seasons. Along the Cascade Valley the land still testifies to the 1868 and 1889 complexes. Cliff notes that today's trails all follow routes that appear on the first maps of Banff. All derive from aboriginal trails, which almost certainly coincided with animal paths. All track along south- and west-facing slopes, those that are the warmest and driest, that shed snow the earliest. They are the slopes most readily burned.

The Banff fire program has tried to replicate that heritage, particularly where the valley narrows against the Palliser Range and thickening trees threaten to strangle what should be a major wildlife corridor. The hillsides are pocked with the strips and splotches of prescribed fires, among the earliest big burns the program attempted. They sprayed the slopes with ping-pong-ball incendiaries. The fires, however, skipped and splashed, and proved less devouring than the program expected.

The various paths invite different strategies for moving our pack stock along. Sometimes we pull them on a short lead. Other times, say, picking our way down a steep or rocky path, the lead is loose, the horse allowed to select its footing. When the trail widens along an old road, the stock may be left to fan out. There may be some nipping and shoving among the group to achieve the proper pecking order, but most accept their place in the sequence. Ribbon is notably effective at marshalling. He cuts off Ziggy when Ziggy tries to hustle ahead, and nudges Hillary along when she dawdles. The only incident occurs when Rocky pulls free along a narrow, sloping trail through lodgepole and dashes up the hill. Ian lets him prance and cavort and when the trail doubles back on the other side of the ridge, we intercept Rocky and reclaim his halter.

<hr>

Flint's Camp is a newer-model cabin. After unpacking, Cliff strikes for the peaks where he believes he has spied a wolf chasing rams. Ian and I hike along a decaying trail that leads to the old cabin, near the headwaters of the Cascade River. Ian is unsure why the old cabin was removed. (The newer site, while less remote, has poorer pasture.) Still, keeping with Banff tradition, the cabin was relocated, not abolished.

There are new fires around the meadows across the stream from Flint's Camp, burned in 2001 much as they were around Scotch Camp. The mountains looming behind harbor the scars of many predecessors. In 1936 the largest fire of Banff's biggest-known fire season ripped across its slopes and tore through Cuthead junction into the splayed valleys beyond. Probably the blaze began as an escaped campfire near Badger Pass. A record drought, powerful winds, a large expanse of forest—the fire proved unstoppable. Only the 1940 season came close to approaching it for scale. Since then wildfires have virtually ceased. The only fires of any power are those that the park's fire officers kindle.

Later, Hillary manages to knock apart the small log fence by the stream and fords the waters into Flint's Park itself. The others hesitate; for some reason they are reluctant to sprint wild and free to join her. Stephen spots the breach and we entice Hillary, who seems unsure of what she has done and why, to return to the corral. Cliff and Ian hammer the logs back in place. Later, because the small corral lacks extensive pastures such as those at Scotch and Windy, Ian serves an extra ration of oats.

<hr>

The trek to Stony Cabin puts us squarely back into Cascade Valley and eventually over the gravel road that once joined Banff townsite with Ya Ha Tinda. We pause for lunch at the historic Elk Trap Meadow, the site of an enormous log structure for corralling elk, now decaying. The park has burned lightly around the structure to protect it from wildfire. Broadcast burning on the mountain slopes commenced in 1986, and was

repeated in 1990. In 1992 wolves returned. They promptly slashed the elk
population from 250 to perhaps 20–40. Part of the pack has followed
the retreating elk into Banff townsite itself. That is where the park's fire
program must go, and where we are headed as well.

───────────────

All week the weather has been unsettled. We have faced stiff winds and
low clouds, snow and drizzle over the passes, blistering sun that burned
our hands and necks, then wind, scudding clouds, more dazzling sun, and
now the sky boils and threatens a thunderstorm. We get the horses safely
pastured at Stony Cabin.

Cliff, Ian, and I hike to a campsite a bit up the valley. The park burned
a small side valley last fall. We search for aspen suckers and elk pellets,
and Cliff and Ian recall how, on this same site, a few years previous, a
handful of scientists had faced off in what became mockingly known as
the Battle of the Stoney Creek Outhouse Meadow and argued, as only
academics can, about whether the fire regimes of Banff were anthropogenic or natural. The brouhaha knotted several themes together, some
scientific, some cultural.

The science thesis was that Canada's grand fire history obeyed climatic
forces and that humans could no more alter fire's outcomes than they
could the ebb and flow of ice ages. Fire management had no fundamental
impact on fire's geography. There was no point in fighting fires, there was
no point in lighting them, there was no point in fussing over fuels. The
statistical outcome would be identical. There was no point in having a
fire program at all except to save buildings from wildfire. This bold claim,
backed by graphs, merged with biocentric beliefs to argue that nature, and
nature alone, should be left to run its affairs. Whatever people did was
either irrelevant or meddlesome.

The cultural argument, while unstated, ran in parallel. The political
entity called Canada is itself an institutional response to outside forces
beyond its control, of which the first and mightiest is climate. Climate
change, moreover, is an apt model for other exogenous forces—imperial,
political, economic, cultural—which Canada has to endlessly accommodate. It was possible to imagine Canada as a confederation of convenience
arrayed to protect its population against global or continental powers over

which it has little direct influence. The institutional turmoil that so characterizes its bureaucracies and its ever-looser union is, in this sense, only
a token of Canadian adaptations against what it cannot change.

Banff saw matters differently. In particular, Cliff and his colleagues
saw the hand of humanity widely sculpting Banff's historic landscapes
and they believed that removing that not-so-invisible hand from nature's
economy could unravel Banff's biota, like a sloppy diamond hitch ready
to dump boxes along a trail. The Banff model suggested that Canada was
not merely a shelter for survival but a positive act of human imagination
and social will.

The fire history controversy nearly split Parks Canada along ideolog-
ical lines, and it dovetailed into another crisis that developed at Banff
shortly after Cliff White returned. The explosive urban growth around
Banff townsite—almost 10 percent per year—threatened to overwhelm
Bow Valley. Unless that immense corridor remained open to wildlife,
the Banff fauna could not hope to adapt and would be driven into deep-
mountain holding pens. The park could not have both. Cliffie hurled
himself into the task with characteristic élan. The environmental crisis in
the Bow Valley could not be separated from politics, and neither could
extricate itself from the ecology of fire. Wildfires would come when they
weren't wanted, and controlled burns would be denied when they were
needed. Banff's fire program moved out of the backcountry.

The 1993 Sawback prescribed burn to the west of Banff townsite alarmed
and rankled many observers. It was big, near, and visible in ways that
backcountry fires in the Lower Panther or Flint's Park were not. The idea
was to burn hugely: to burn broadly enough that aspen suckers could
proliferate more abundantly than the resident elk could crop them off,
and to begin pruning the fire hazards of the Bow Valley. Instead, the
project aroused public ire and bumper stickers that read "Fire a Warden,
Save a Forest." The fire flushed into the open the various environmental
maladies of Bow Valley.

The arguments for doing nothing were attractive. Administrators
had enough controversy on their hands without worrying about possible
escape fires and valley-clogging smoke and carping from critics who dis-

missed the burns as alternately meaningless and dangerous. It was also a way to pen the restless Cliffie who had become intimately involved—intrusively so, to critics—in the massive Bow Valley Study, a comprehensive inquiry into the multiple ecological pathologies of Banff townsite and its environs. Tension mounted; the fire program was slammed shut. Then, in December 1995, Cliff was dispatched into exile. The park was reorganizing, he was told. There was no longer any place for him in a supervisory role. He needed to disappear.

In January 1997 he enrolled in a PhD program at the University of British Columbia. His thesis topic: wolves, elk, aspen, and fire in Banff. He graduated in June 2000. By then the forces that had pushed him out had reconsidered. The Bow Valley Study had ended, its raw nerves healed over, but concluded basically along the lines he had predicted it had to, a paradigm for the Canada-wide ecological integrity panel that succeeded it. The fire program had roared back, now a paragon of Parks Canada's commitment to putting policy into practice. Cliff returned to Banff, still without an administrative post, something of a gadfly, or better, a catalyst, not unlike the fires he had urged with such dash and conviction.

All afternoon the sky had darkened until lightning announced a climax and the downpour washed over Stoney Cabin. We could hear thunder echoing off the mountains.

Cliff seems unconcerned about possible fires the storm might kindle. Banff is typical of the east-slope Rocky Mountains in that, while lightning pounds the west side, kindling fires abundantly in Kootenay and Glacier Parks, it only rarely does so in Jasper and Banff. A few fires do start, and from time to time one spills over the divide, driven by chinook winds. But Cliff's study of Banff fire history had concluded that lightning accounted for only a tiny percentage of fires, although that percentage has grown as the overall population of fires has shrunk. From 1880 to 1980, lightning kindled only 7 out of the 53 fires that exceeded 40 ha. The more startling statistic was that the ratio of human-caused to lightning fires had been 25:6 from 1880 to 1940, but was only 1:1 thereafter. The common wisdom was that there were far fewer fires today because people had suppressed them. More reasonably, the fires had disappeared because people no longer lit them, whether deliberately or accidentally. People had set virtually all the giant fires of Banff's preceding two centuries. The 1936 burn spread from a campfire. The 1889 fires were sparked by the railroad. The 1929 Cuthead fire escaped from "careless campers."

The storm passes to the south, between Stoney Camp and Banff townsite, where the Cascade Valley joins the Bow. There, on the southeast-facing flanks of the Cascade Mountains, stands a large patch of lodgepole pine. It has not burned since 1701. No one knows why.

―――――――――――――

Banff townsite lies a scant eight miles distant, where the widening Cascade Valley meets Bow Valley. We shun the direct route down Cascade—already we have encountered day recreationists. Instead we veer westward and curve around the Cascade Range, where a creek breaches it and where in 2001 some prescribed fires attempted to burn off the south-facing slopes. But we can delay the exurban penumbra no longer. Its tentacles reach deep up the valleys.

We pass signs warning of mountain bikes. We pass the Mount Norquay ski lift. We pass day hikers and riders. We hear helicopters overhead; the first is on a fire reconnaissance after the storm (they find nothing), but others follow. We pass underground utility corridors for gas and water. We come to the Trans-Canada Highway. There are cars, trucks, SUVs, cars with sirens chasing other cars, trucks hauling more cars. There are planes and helos overhead. There are lights and power lines. We pass underneath the divided highway through a wildlife tunnel. We cross the tracks of the Canadian Pacific Railway. We come to another highway, and press electric buttons to trigger traffic lights to halt the rush long enough for us to lead our stock over the asphalt.

We are traveling along Banff's greatest corridor. Everything else in the park bends to its flow. It is, not incidentally, a route of fire—of internal fire and fossil fuel, yes, but still a passage shaped by controlled combustion in the hands of humans. We have crossed the threshold into the world of industrial fire. Not far from Banff townsite, but well within the park boundaries to the east, are the remains of Anthracite, a coal-mining town, a reminder that the coal and steam in the form of the Canadian Pacific initially created the park, Canada's first.

The two Cliffs, father and son, discuss the changes the past two decades have wrought. They are both native Banffians, the second and third generation to be born in Banff, and the fourth generation to reside there. The family patriarch, Dave White, arrived with the Canadian Pacific in 1885 and stayed. One member with artistic ambitions, Peter White, changed the spelling of his surname to read Whyte, married well, and bequeathed the Whyte Museum. Their family history is inextricably entangled with that of Banff.

Cliff Sr. has worked at most jobs Banff offers—mountain ski guide, road maintenance cat skinner, gas station operator, ski area manager for the Sunshine ski resort. He knows the frustrations of living under colonial rule from Ottawa, as Banff townsite did until home rule was granted in 1988 and it could manage for itself the daily business of a town. He understands the travails of a highly seasonal economy dependent on tourism and government edicts and appreciates the extent to which those commercial services make the park accessible to the Canadian public. Yet he also thrills to the park's splendors. He remains, at 71, a spry outdoorsman; two sons have entered the warden service. He and Cliffie discuss plans to meet with their extended relatives in the evening. The park and its township, he knows, can neither exist without the other.

The town resides like a small atoll amid a rising sea of conifers. The earliest photos of Bow Valley show a landscape of meadows with scattered trees, riparian woodlands, and patchy wetlands. Today, save where carved out by buildings and highways, it bristles with close-packed pine. A large crown fire—and that is the only form a large fire would take here—would blow through the town like pine pollen. Of all the landscapes in Banff, its signature city boasts the highest values and the greatest risk from wildfire.

The problem is that town and country don't mix, rather like a pack string meeting the Trans-Canada Highway. If Banff were a town, it could create an urban fire service. If it were a cluster of backcountry camps, it could adapt to wildland fire. But it is neither, and both. Worse, its metastasizing lodgepole pine have reached their prime age to burn as a crown fire. The most probable kind of fire is the one that will inflict the most damage. How to kindle prescribed burns to keep the Bow as a wildlife corridor while protecting the town from wildfires is an unenviable task that involves clearing extensive fireguards and selectively burning.

Cliff and Ian point out where this has been done at Noquay, Carrot, and Sulfur, though it is difficult to see much through the dense thicket of trees. It is harder still to imagine the hybrid combustion regime that would result. Urban folk distrust open flame and detest smoke. But, against all forecasts, Banff townsite has learned to accommodate passing wolves and cougars. (One cougar has even killed a resident.) They might, over time, learn to accept an equivalent dose of free-burning fire. The difference is this: they could kill off the wolves, grizzlies, and cougars if they wished. They can't kill off fire. Sooner or later fire will swarm to the town's borders. In 1841 a fire filled the Spray Valley, which joins the Bow exactly at Banff townsite, with wall-to-wall flame.

Across the highway we ride through a lovely meadow at Whiskey Creek. Wolves have appeared sporadically here, warily threading under the highways as they might cross a frozen river; bears have not yet found their way through the maze. The aspen show hard wear from elk. The grove has not burned for decades. We ride on, across the pale of settlement. There are roads, shops, hotels, even a cemetery. At last, we come to a vast complex of corrals and barns, a scene as incongruous as a swidden patch in metropolitan Toronto.

Whatever threat fire might pose to the town, the greater threat is the one the town poses for fire. Banff town and Banff Park headquarters are the prime movers of this biota's fire factory. Information and institutions rival drought and forest structure as molders of flame's dynamics. Humanity's control is hardly total: lightning will persist, and some wildlands will burn, although perhaps in ways very different from those they have known for millennia. Yet today the fire regimes of Banff are indisputably shaped by humanity. The park decides whether and how fires will be suppressed, whether and how lightning fires will be allowed to burn, whether and how prescribed fires will be kindled. The park hired Cliff, the park nearly fired him, the park reinstated him. What Cliff and Ian and Parks Canada en masse have realized, moreover, is that these circumstances did not begin with the Canadian Pacific Railroad and the

Banff Springs Hotel. Rather, humanity has, since the retreat of the ice, structured the park's fire landscape as fully as wolves have influenced its dynamics of elk and aspen.

It is the ultimate top-down model of ecology. Among the many megafauna of Banff, one shapes its surroundings not simply by digging or hunting or chewing trees but by burning or not burning. Hominins are the fire creatures: they are, through fire, the keystone species that will, for good or ill, wisely or idiotically, intentionally or accidentally, catalyze a world that the other creatures must inhabit. Wolves, grizzlies, cougars, elk, mountain goats, bighorn sheep—each has unique qualities, but to some degree another can substitute. If a niche opens, some other animals will enter it. But not with fire. There is no other biological source of ignition than humanity. Life created oxygen, life created fuel. What life could not do is kindle, or it could not until *Homo* arrived, and allowed the biosphere to very nearly close the cycle of burning. If humans fail, there is no other creature to do it for them.

We unsaddle and unpack for the last time. Horse trailers will haul some of the stock back to Ya Ha Tinda. A van and pickup wait to take us to park headquarters where, in truth, our trek has trended from the time it began, for our journey from the Red Deer Valley to the Bow has been a journey through time as well as space. The ontogeny of our travels has recapitulated the phylogenic history of humanity as an ecological agent in Banff. Along the Red Deer aboriginal North Americans walked, hunted, foraged, and burned. Along the Bow today, contemporary Canadians drive, observe wildlife, snap photos, and burn—or not. They must decide whether to apply or withhold fire and in what forms and to what ends. They remain the predator of predators. They have, as part of their genetic inheritance, a capacity beyond fangs, talons, claws, and smell, a power greater than the tireless lope of a wolf or the brute strength of a grizzly. They can start and stop fire. If Cliff White and his cohorts are anywhere near correct in their reading of that land and its history, fire is the keystone catalyst for its future, whatever they and nature decide that future might be.

A JOURNEY TO THE SOURCE

This is where fire goes to die. If you think fire is difficult to live with, try ice.

I HAD NOT INTENDED to go to the end of nowhere. Even by Antarctic standards Dome C was the back of beyond. But one afternoon, while fumbling with gear in a McMurdo warehouse, I overheard an allusion to "the source regions." The technicians used the term casually, discussing what goods would be shipped to a remote camp well over the Transantarctics. But in my mind the phrase sparked an epiphany of symbolism. The source region. Here was the geographic place at which the East Antarctic ice sheets gathered and then flowed outward. Here was a place that took nothing from elsewhere save fugitive water vapor and turned it into Ice. I had come here to understand Antarctica, by whatever means I could. Surely that quest demanded a journey to the source, for it must certainly contain the essence of the Ice. So after a trip to the National Science Foundation chalet where I pleaded my case, and then a layover at Pole, adapting to high elevation, I stepped off an LC-130 Hercules ("The Antarctic Queen") on a dazzling 1981 New Year's Eve at Dome C and found myself at the end of the world.

===

Antarctica is a place only an intellectual could love. The further one moves into the interior, away from the coast and storms and marine life that tenuously valence with the Earth, the more dominant the ice and the more extraterrestrial the surroundings. The commonsense perspective of ordinary people is that there is "nothing there," and they are almost right. Even scientists in keen pursuit of data, precious by being rare, our age's equivalent to the spices and bullion that inflamed early explorers, found Dome C extreme. The rumor soon spread, originating from a knot of geophysics grad students from Wisconsin, that we were not in Antarctica at all but had been secretly drugged on the plane and taken to a prison camp in Minnesota.

The occupation was temporary—how could it be otherwise? Here, where the ice thickened on a continental scale, was an ideal setting for deep-core drilling. Exploratory flights had resulted in a crippled LC-130, when a taxing plane dipped a wing into sastrugi; it caught and cracked. Another ski-equipped Hercules was dispatched to remove the crew, only to add to the crisis when a jet-assist takeoff bottle broke loose on takeoff and ripped through a wing. A third flight rescued the stranded. The next year a temporary camp of Jamesway huts arose while crews repaired and flew out the planes. The Jamesways became the core of an ice-prospecting camp. A small cadre of French glaciologists sat in their self-proclaimed *cage aux folles* and sank their coring shafts. Smaller cliques of Americans hand drilled for shallow cores and blasted for seismic profiles that revealed an ice sheet roughly 14,500 feet deep. A nosecone from one of the former LC-130s, like a cannon on a village green, greeted newcomers.

In all there were a dozen of us—the French, the American teams, a cook, a couple of navy mechanics. But there was little that one might consider a society. There was nothing from the outside world by which to order the inside world at Dome C. Nothing in nature, nothing in culture, only the fortnightly visit by a resupply Hercules. As the sun slowly spun above the horizon, teams came and went as their work called, or they felt an urge. There was no common dining, no collective experience, nothing that anyone had to do at any one time. The sole exception was the arrival of new movies with each resupply flight. These were watched obsessively until the cache was exhausted, at which point Dome C's social order again dissolved.

A naïve observer might rejoice in the near-absolute freedom allowed by a near-absolute abolition of mandatory order. But that nominal freedom is only another name for anomie. Freedom is relative: it requires coercion of various sorts in order to have meaning. At Dome C there was nothing to rebel against. You could do whatever you wished. The catch was, there was almost nothing to do. Those with complex projects, originating from the outside, survived better than those without. But like food brought in, the project exhausted itself with use, and as the Ice inevitably ablated these away, their practitioners survived by departing. No one lived at Dome C. Those who stayed longest sank into various pathologies. The Big Eye, or insomnia—I went on a 24-hour cycle of wakening followed by 12 hours of sleep. The Long Eye, aptly defined as a 12-foot stare in a 10-foot room. You slow down. With little to stimulate you, there is no reason to busy yourself. Stay for long, and a state of semihibernation sets in. Stay too long, and you will find yourself dissolved in a psychic whiteout with the Ice.

The Barrier ice is a term derived from James Ross's expedition to the eponymous Ross Sea, and originally it referred to the (also eponymous) ice shelf—that immense floating delta of glacial ice, as vast as Texas or France (depending on your reference system)—that fronted the sea like the White Cliffs of Dover. The Source regions go the Barrier ice shelf one step farther as a frontier. The ice shelves collect the flow of glaciers from east and west Antarctica; the source regions gather nothing. They simply accept particles of ice.

Across the Barrier life ends. It can exist for skuas only by flying over and back, or for humans, by sledging out and returning. This is a wholly abiotic environs. Its energy flux is ever negative, it lacks flowing water, it is void of nutrients. There are not even rocks that might, in principle, disintegrate into a substratum of raw elements. There is one molecule: hydrogen dioxide. There is one unblinking scene: a sheet of ice. Oxygen abounds, so one can breathe, but there is nothing else to support organisms. People can live only through umbilical cords and IV drips to a sustaining society well beyond the reach of the Ice. Left to itself, life feeds off itself, and then shrivels.

The ice sheets are equally acultural. There is no basis beyond the Barrier for norms of social behavior or sources of knowledge other than what further connections with the outside world can bring to bear. There is no prospect for an Antarctic explorer to recapitulate Alexander von Humboldt's ecstatic contact with the Venezuelan jungle, picking up one new specimen, only to drop it for another, and another. There is no engagement with indigenous peoples as guides, interpreters, and collectors, or any means to go native and immerse oneself into another moral universe. To reach the North Pole Robert Peary adapted Inuit sledges and dogs and relied on native sledgers; to reach the South Pole, Robert Scott's Polar Party pulled their own sledges.

With ever dwindling social entities—the sledging party is less synecdoche than symbol—there are few opportunities for the kind of contrast and conflict that drives literature, and scant scenes ready for the visual arts. When he learned that he would not join Scott's Polar Party, Herbert Ponting, the expedition photographer, decided it was probably for the best. Other than portraitures, there would be "nothing" to photograph. Beyond the Barrier even intellectuals struggled to find sustenance; they relied instead on the elaborate baggage they brought with them to create comparison, contrast, and context, without which their minds would find nothing with which to grasp. Antarctica became known less for what it was than for what it was not.

The classic photography of Antarctica is thus not a photography of ice but of Barriers—the edge, where ice meets sea, rock, and sky, where life pokes and flaps and swims against the border, where things move and sounds echo. Beyond the Barrier lies a nature like a modernist painting: abstract, conceptual, minimal. The literature of Antarctica is likewise not a record of social exchange or innovation or surprise but a chronicle of diaries and soliloquies, the self withdrawn, drafting from its own reserves for its sustenance, like a camel on its hump. It thrives on whatever it has stocked from elsewhere. Over and again, literature recycles the same stories. Those Antarctic archetypes are not only all that exist but perhaps all that can exist. The opportunities for endless variants simply aren't present. Instead, imaginative literature turns to fantasy and science fiction, a realm beyond reason and empiricism, a dominion as intrinsically blank as the ice sheet. What the ice has done to landscape, it does to society, and hence to the social imagination of art. All that remains are ideas.

When Richard Byrd attempted to live alone for a winter, crossing the Barrier during his second expedition, creating 150 miles away from his Little America base a kind of Antarctic Walden, the experiment went bad. He became hallucinatory, suffering a kind of dementia, before the experience nearly turned lethal. The official explanation was carbon monoxide poisoning, but the truer answer might be the folly of trying to simplify existence amid what was already so simple as to belong on a moon of Saturn. Survival required exactly the opposite strategy. It demanded complexity, sensory overload, a Victorian museum cluttered with bric-a-brac rather than a spare, wooden shack. Otherwise, society breaks down to a self.

But it can go further. It did for me. Unlike those I traveled with, I had no specific task, no experiments to perform, no drill to sink, no rocks to collect. I did not arrive, do my job, and depart. I stayed for a full season, attaching myself like a lamprey to whatever larger enterprise I could find. There were often occasions where I simply waited. I had no shield, save what I brought by way of books and ideas, between me and the Ice. I had no forced busyness to insulate me against the sapping cold. There were no intrinsic stimulants from the outside. Gradually the realization sank in that Antarctica did not offer a unique experience so much as the experience of having the familiar world removed. It was a place of things that should be there and weren't. It lacked that quantum of complexity without which culture cannot work. Living there was a process of social reductionism that led to a cultural numbing, a mental hypothermia. Antarctic was the sum of its losses.

Even so, even as the ice thickens in much of Antarctica, there are borders around. The shelf slides off mountain and into sea. Glaciers grind through rocky passes. The grand ice sheets, east and west, have their geographic frontiers. But the Source does not. There is only that sharp trace between ice and sky. Otherwise, ice meets ice. What makes Antarctica what it is, its ice, is here distilled into ice alone, and what makes the Source region unique, both sublime and nihilistic, is that here ice works self-referentially even upon ice. Here in the center nothing holds because there is nothing to do the holding and nothing to hold. There is no Barrier: there is almost nothing at all. At the Source it becomes clear just what the ice does. It simplifies. It takes. It reduces. It reflects.

That geosocial nihilism ablates away anyone exposed to it for any length of time. The protective shields of food, portable energy, memories, and tasks brought from elsewhere gradually sublimate. The wise leave before their imported stocks are exhausted. The foolish or the unlucky must watch their self exposed only to the ice, which is to say, to attrition. There is not enough on site to generate those contrasts that allow ideas to arc between them. The ultimate Antarctic experience is not a scene at all but a whiteout. Yet the Self can only exist—can only be felt and known—in contrast to an Other. At the Source there is no Other. There are no other creatures, no other environs, no other emblems of a world beyond. There is no basis for meaning. There is only Ice.

Consider the geographic facts. Dome C is an infinitesimal rise in the East Antarctic plateau, atop 14,500 feet of ice that spreads for hundreds of miles. There is ice and little else. This is the most singular environment on Earth, a synthesis of the huge with the simple such that the continent is reduced to a solitary mineral broader than Australia and higher than Mount Whitney. Space and time dissolve. The cycle of days and those of seasons collapse into a single spiral. The energy budget is always negative; none during the dark season, reflected away during the light. There is no life. There is nothing to live on. Here is Dante's imagined innermost circle of hell as an inferno of ice. Here is the Earth's underworld.

It is a scene of absences and abstractions. There is no color, no movement, no sound. There are no mountains, valleys, rivers, shores; no forests, prairies, tide pools, corn and cotton fields, sunbaked deserts; no hurricanes, no floods, no earthquakes, no fires. The only contrast is between an ice-massed land and an ice-saturated sky. The descending ice that links them—the ultimate source of the dome—has the purity of triple-distilled water. Yet it too, as with everything else, simplifies into its most primordial elements, as snowflakes crumble and fall as an icy dust. There is no center and no edge. There is no near or far; no east or west; no real here or there; no Other, and during a whiteout, no self. Words, too, shrink and freeze, as language and ideas shrivel into monosyllables: ice, snow, dark, sky, blue, star, cloud, white, wind, moon, light, flake, cold. Here is the end of the world.

Consider what that does to experience, to mind, to self. Like the flakes disintegrated into slivers of crystal, a mounding of ice dust, the self disaggregates. Your self is not an essence, but the compounding sum of your connections, like snowflakes elaborating uniquely. At Dome C every particle of ice is identical.

———

When I returned to McMurdo, I began regaining my senses, and when I reached Tierra del Fuego, I savored smell and sound in a landscape lush with life and color, though one that the rest of Earth had long dismissed as impauperate. I craved contrasts—the darkness that I had not known for three months, the flowing water I had been denied, the aromas of plants. At home I rejoiced in every sensation. The most mundane object and sentiment felt keener. The rustle of wind through trees. The odor of animals. The bustle of an airport. The taste of tomato. The touch of family. Even, since it was the boreal winter, the sight of snowflakes, like filigree webs, on a windowsill.

But I carried away also a memory that made Dome C the Ineffable as palpable as a granite outcrop. After nothing, everything seems sublime.

NOTE ON SOURCES

S INCE THIS IS NOT a regionally focused volume, there are no common background references I have relied on. Each essay contains
those sources that pertain to it. Otherwise, the primary inspiration
has come, as throughout the series, by personal observation, by travels, and
by conversations with practitioners and researchers.

NOTES

SLOUCHING TOWARD GATLINBURG

1. Charles W. Lafon, Adam T. Naito, Henri D Grissino-Mayer, Sally P. Horn, and Thomas A. Waldrop, *Fire History of the Appalachian Region: A Review and Synthesis*, General Technical Report SRS-219, U.S. Forest Service, January 2017, 3.

2. Lafon et al., *Fire History*, 3. See also William T. Flatley, et al., "Changing Fire Regimes and Old-Growth Forest Succession Along a Topographic Gradient in the Great Smoky Mountains," *Forest Ecology and Management* 350 (2015): 96–106.

3. For background on the fire, see "Analyzing the Fire That Burned into Gatlinburg," *Wildfire Today*, December 5, 2016, http://wildfiretoday.com /2016/12/05/analyzing-the-fire-that-burned-into-Gatlinburg. Also, see National Park Service, Smokey Mountain National Park, *Chimney Tops 2 Fire Summary, 12/12/2016*. The *Knox Mercury* analyzed the catastrophe as follows: http://www.knoxmercury.com/2016/12/07/gatlinburgs -inferno-started-spread-needs-happen-next/. For a later informal review by the National Park Service, see "NPS Official Talks About the Wildfire That Burned into Gatlinburg," *Wildfire Today*, June 13, 2017, http:// wildfiretoday.com/?s=nps+official+talks+gatlinburg.

4. See "Charges Dropped Against 2 Teens Initially Thought to Have Started Wildfire That Burned into Gatlinburg," *Wildfire Today*, June 30, 2017, http://wildfiretoday.com/2017/06/30/charges-dropped-against-2 -teens-initially-thought-to-have-started-wildfire-that-burned-into -gatlinburg/.

SPOTTING PACK RATS

1. Personal communication from Mike Rogers, 2011.
2. Christopher Doyle, "Little Marsupial Diggers May Hold Key to Preventing Bushfires," ABC Environment, June 30, 2014, http://www.radio australia.net.au/chinese/2014-07-01/little-marsupial-diggers-may-hold -key-to-preventing-bushfires/1335840.
3. R. I. Yeaton, "Porcupines, Fires and the Dynamics of the Tree Layer of the *Burkea Africana* Savanna," *Journal of Ecology* 76 (1988), 1017–29.
4. Personal observation, Alpine, Arizona.

PROMETHEUS SHRUGGED

1. Jerry Williams, "Mega-Fires and the Urgency to Re-Evaluate Wildfire Protection Strategies through a Land Management Prism," unpublished paper presented to the *Exploring the Mega-Fire Reality Conference*, Tallahassee, Florida, November 2011.

UNTAMED ART

1. Biographical information from Svetlana Semenova, *Ocharovan Uralom: Zhizn' i tvorchestvo A. K. Denisova-Ural'skogo [Enchanted by the Urals]* (Sverdlovsk: Sred.-Ural. kn. izd-vo, 1978). I am indebted to Irina Petrova James for translating and interpreting the relevant passages, which has made my synopsis possible.

   Aleksei's birth year is disputed; some claim it to be 1863 or 1864. I follow Semenova.
2. Russia submitted over 600 works of art, representing the collections of ten art societies, to the fair. Denisov-Uralsky's silver medal was one of 12 awarded. See Mark Bennitt, ed., *History of the Louisiana Purchase Exposition* (New York: Arno Press, 1976; reprint of 1905 edition), 515, and especially Robert C. Williams, *Russian Art and American Money 1900–1942* (Cambridge, Mass.: Harvard University Press, 1980), 42–82; on the Wanderers and Denisov-Uralsky, see Williams, 43–44. A biographical sketch for the exposition listed membership in the following Russian art organizations: Exhibitors of the St. Petersburg Society of Artists; Spring Exhibitors of the Imperial Academy of Arts; Society of Amateurs of the Fine Arts in Ekaterinberg; St. Petersburg Art Society; from

Letter of Halsey Ives to R. T. F. Harding, May 2, 1905, Halsey Ives Archives, St. Louis Art Museum.

3.   Quotes are from Semenova and reflect her judgment based on letters sent from Finland, *Enchanted*, 103–12.

4.   I rely on Williams, *Russian Art and American Money*, for the details of this story; quote from 49.

5.   All details and quotes are again from Williams, *Russian Art and American Money*, 49–52. On Grunwaldt's 1891 experience, in which the French organizer went bankrupt, see George F. Kennan, *The Fateful Alliance: France, Russia, and the Coming of the First World War* (New York: Pantheon Books, 1984), 89–90. On Japan's acquisition of Russia's exhibit space, see "Extensive Exhibit," *Washington Post*, January 9, 1905, 11.

6.   Description from *Russia: Fine Arts*, Catalogue of the Russian exhibition to the World's Fair, Saint Louis, USA, 1904, Louisiana Purchase Exposition (St. Louis, 1904), 43, 59. Appraisal of painting from letter from Halsey Ives to R. T. F. Harding, May 2, 1905, Ives Collection, St. Louis Art Museum Archives.

7.   On plans for Portland, see "Extensive Exhibit," *Washington Post*, January 9, 1905, 11.

8.   On Grunwaldt's decline, see *New York Times*, May 22, 1906, 6; and on his lawsuit, *New York Times*, August 9, 1907, 7, and June 12, 1909, 9. Estimated cost in today's dollars from the online calculator posted by the U.S. Bureau of Labor Statistics. That record only reaches back to 1913.

9.   The sales are described in the *New York Times*, March 7 (p. 9) and March 8 (p. 3), and the confiscation on March 10 (p. 6). The *Times* reported that it "appeared that the works already sold had been taken out of bond and the tariff on them paid." Quote on Kowalsky from Williams, *Russian Art and American Money*, 61.

10.   Converted to 2018 dollars, Havens's acquisition cost him $1,024,473, a substantial sum, but only about $1,888 per painting, a bargain for the cream of an exhibition.

11.   Williams, *Russian Art and American Money*, 79–80.

12.   On Busch's purchase, see letter from Halsey Ives to R. H. Harding, February 15, 1905, Halsey Ives Collection, St. Louis Art Museum. I'm indebted to archivist Norma Sindelar for securing this vital document. On Busch's role in the exposition, see Bennitt, *History of the Louisiana Purchase Exposition*; a photo and caption summary are available on p. 85.

13. See letter of R. H. Harding to Halsey Ives, February 11, 1905, Halsey Ives Collection, St. Louis Art Museum.

14. Quote from George Adams, "Today's Talk," *Gettysburg Times*, June 22, 1938, 4.

15. See "Important Notice," *Cedar Rapids Evening Gazette*, November 9, 1909, 3. U.S. Forest Service historic photo collection, accession number 454356.

16. On Busch's motives, I rely on comments from Robert Williams. No official reason was given other than a vague gesture of friendship, and Anheuser-Busch has offered no formal explanation.

17. Dobrynin's comments in letter from James McCargar, National Endowment for the Humanities, to Robert C. Williams, July 30, 1979, p. 3 (copy courtesy of Robert Williams). Information regarding the Russian scene in an email from Olga Klimenskaya, Curator of Russian Art for 18th–20th Centuries, Perm State Art Galleries, to Tatiana Volokitina, and forwarded to author on June 8, 2007.

18. I am indebted to Ludmila Boudrina for relaying to me the latest verse in this sad saga and for sending me a copy of the catalogue raisonné for the exhibit.

IN THE LINE OF FIRE

1. Reports from Global Fire Monitoring Center, which aggregated news reports and satellite imagery. See "Forest Fires in South and North Korea: 08 April 2005," Global Fire Monitoring Center, accessed October 12, 2015, http://www.fire.uni-freiburg.de/GFMCnew/2005/04/0408/20050408_korea.htm. To access use "fire" as the identification and "1" as the password.

2. Background on agriculture from Michael J. Seth, *A History of Korea: From Antiquity to the Present* (Lanham, Md.: Rowman and Littlefield, 2011), and Michael E. Robinson, *Korea's Twentieth-Century Odyssey: A Short History* (Honolulu: University of Hawaii Press, 2007).

3. A concise survey of Korea's environmental evolution played against modern politics is Lisa M. Brady, "Life in the DMZ: Turning a Diplomatic Failure into an Environmental Success," *Diplomatic History* 32, no. 4 (2008): 585–611.

4. An excellent summary of these developments, and the source of my figures, is Jino Kwon et al., *Forest Landscape Restoration Success, Emerging Challenges, and Future Direction in the Republic of Korea* (Korean Forest

Research Institute, 2014). The fire essence is distilled into a brochure also by the Korean Forest Research Institute, *Forest Ecosystem Change Since 1996 Wildfire in Korea*, n.p., n.d.

5.   Korean Forest Research Institute, *Lost Landscape in Forest Wildfire: 20 Years Changes at Eastern Coast of Korea* (Korean Forest Research Institute, n.d.), p. 8. A nice pocket-sized summary is available in Soung-Ryoul Ryu, English editor, *Post-Fire Restoration to Establish a Healthy and Sustainable Forest Ecosystem* (Korea Forest Research Institute, 2010).

6.   On the 2000 fires see Kwon et al., *Forest Landscape Restoration*, 67–68, and Global Fire Monitoring Center, accessed October 21, 2015, http://www.fire.uni-freiburg.de/current/archive/kr/2000/04/kr_04172000.htm.

INDEX

ABOUT THE AUTHOR

Stephen J. Pyne is a professor in the School of Life Sciences, Arizona State University, and a former North Rim Longshot. He is the author of over 30 books, most recently *Between Two Fires: A Fire History of Contemporary America* and *To the Last Smoke*, a series of regional fire surveys. He lives in Queen Creek, Arizona.